# ECOACCIDENTS

# NATO CONFERENCE SERIES

| I | Ecology |
|---|---|
| II | Systems Science |
| III | Human Factors |
| IV | Marine Sciences |
| V | Air–Sea Interactions |
| VI | Materials Science |

## I ECOLOGY

# ECOACCIDENTS

Edited by
## John Cairns, Jr.
Virginia Polytechnic Institute and State University
Blacksburg, Virginia

Published in cooperation with NATO Scientific Affairs Division

**PLENUM PRESS · NEW YORK AND LONDON**

Library of Congress Cataloging in Publication Data

NATO Advanced Research Workshop on Ecoaccidents (1983: Noordwijkerhout, Netherlands)
  Ecoaccidents.

  (NATO conference series. I, Ecology; v. 11)
  "Proceedings of a NATO Advanced Research Workshop on Ecoaccidents, held September 19–25, 1983, in Nordwijkerhout, the Netherlands"—T.p. verso.
  Bibliography: p.
  Includes index.
  1. Water—Pollution—Accidents—Congresses. 2. Water—Pollution—Environmental aspects—Congresses. 3. Environmental chemistry—Congresses. I. Cairns, John, 1923–    . II. North Atlantic Treaty Organization. Scientific Affairs Division. III. Title. IV. Series.
  TD419.5.N38  1983                        363.7′3                        85-28322
  ISBN-13: 978-1-4615-9452-9      e-ISBN-13: 978-1-4615-9450-5

  DOI: 10.1007/978-1-4615-9450-5

Proceedings of a NATO Advanced Research Workshop on Ecoaccidents, held September 19–25, 1983, in Noordwijkerhout, The Netherlands

© 1985 Plenum Press, New York
Softcover reprint of the hardcover 1st edition 1985

A Division of Plenum Publishing Corporation
233 Spring Street, New York, N.Y. 10013

PREFACE

Dr. Hueck explains in the "History and Development of the
NATO Advanced Research Workshop" some ideas on the genesis of
the Workshop.  The reader should understand that this volume is
a preliminary analysis of a problem that has not received much
attention.  The case histories, for example, are illustrative
because one volume cannot include all histories, or even those
in Europe.  Consequently, case histories are lacking on many
parts of the world, including central Europe and North America,
not because these are unimportant or less disastrous ecologically
than the case histories covered, but because selecting only a few
illustrative cases was possible.  The geographic and ecological
coverage was originally much broader, but, unfortunately, some
participants were unable to attend and this resulted in an even
more skewed distribution geographically.  However, participants
generally agreed that the information necessary to analyze the
problem was covered in sufficient depth by those in attendance
and that other case histories would have been interesting, but
would probably not have caused substantive changes in the
conclusions or recommendations.  The Workshop organizers did not
selectively choose people, companies, or countries to the
exclusion of others.  The committee selected a broad array of
people, countries, and industries, and the participants were
those able to come, prepare a manuscript, and meet the other
conditions.  Thus, what appears to be a skewed distribution was
not planned as such, but was the consequence of people being
unable to attend for various reasons.  Also, limited funding
precluded inviting a large number of persons, and withdrawals,
last minute changes, and so on, made it impossible for every
country and industry to be represented.

J. Cairns, Jr.

CONTENTS

## REGULATORY ASPECTS

## SUMMARY

The very important topics of Modelling and Protocols were
included in the original outline for this book and the meeting
on which it is based.  Unfortunately, a serious illness of one
key participant and other factors for other participants
prevented preparation of these chapters for publication.  This
is also true for Regulatory Aspects and Case Histories – the
Workshop included more than the listed participants who could
not, for various reasons, prepare a written document for
publication.

# HISTORY AND DEVELOPMENT OF THE NATO ADVANCED RESEARCH WORKSHOP ON "ECOACCIDENTS"

H.J. Hueck

TNO Division of Technology for Society
Central Laboratory
Delft, The Netherlands

In 1957 the North Atlantic Council of NATO (North Atlantic Treaty Organization) accepted a report stressing the desirability of expanding non-military cooperation between members of the Atlantic Community in science and technology. On that occasion the Council inaugurated a Science Committee, which has since supported a wide range of scientific activities by awarding research grants and fellowships, and sponsoring advanced study institutes, and other scientific meetings. The Science Committee is advised by "Programme Panels," the "Panel on Eco-Sciences" having been instrumental in initiating an "Advanced Research Workshop on Ecoaccidents," with which this publication is concerned.

The idea of convening a meeting on Ecoaccidents was proposed in 1979 by Dr. Hueck, one of the members of the panel. It was pointed out that accidents such as oil spills from shipwrecked tankers offered a unique opportunity of conducting ecotoxicological experiments in real time and in a real environment on a scale that cannot be realized in the laboratory or even in the field.

For the purpose of implementing this idea, Prof.Dr. D.J. Keunen, of Leiden University, The Netherlands, was asked to chair a preparatory meeting at Delft, The Netherlands, in the spring of 1980, where an ad hoc panel of experts discussed the feasibility of organizing a workshop on ecoaccidents, and presented a recommendation to the "Eco Science Panel". The Panel unanimously agreed on the purposes of such a workshop, which was envisaged to take up the following subjects:
-     reviewing appropriate case histories of ecoaccidents
-     comparing laboratory predictions with actual events.
-     analysing the handling of ecoaccidents with a view to gaining a maximum of scientific information from them.

In view of his reputation and experience in hazard evaluation analysis, Prof.Dr. John Cairns, Jr., of the Virginia Polytechnic Institute and State University, USA, was then invited to convene and organize such a workshop.

In October 1982, a second preparatory meeting, again chaired by Prof. D.J. Kuenen, was organized by the NATO Eco Science Panel. The participants continued their discussions of the previous meeting, and concluded that there was a need to convene a workshop on ecoaccidents or other episodic events. Much work has been done on the continuous discharge of pollutants into the environment, but very little on the effects of ecoaccidents, or on assessing the predictive power of models. The meeting agreed that the workshop was to concentrate _if possible_ on the following major topics:

- Evaluation of methods for predicting effects as occuring in ecoaccidents.
- Evaluation of the type of data needed to document effects of ecoaccidents. The evalution was to be based on the analysis of case history studies of ecoaccidents.
- Studying the predictive power of models, and finding ways of improving it.
- Utilizing the information gained from one environmental disaster in abating another.
- Predicting the prospects of the recovery of ecosystems after an environmental disaster.

This ambitious programme was then discussed by the actual workshop, named NATO Advanced Research Workshop on "Ecoaccidents" and held from 19-23 September 1983, at Noordwijkerhout, The Netherlands, under the chairmanship of Prof. Cairns. The findings of the workshop are recorded in the present publication. The participants in the two preparatory meetings and the workshop itself are listed in the appendix.


Delft, December 28, 1983

# HARBOØR TANGE, DENMARK – A COASTAL AREA POLLUTED WITH

# TOXIC ORGANIC CHEMICALS AND MERCURY

Finn Bro-Rasmussen and Kirsten Warnøe

Laboratory of Environmental Science and Ecology
Technical university of Denmark, DK-2800 Lyngby
&
Danish National Agency of Environmental Protection
Strandgade, DK-1401 Copenhagen, Denmark

## INTRODUCTION

Much public concern has been created by the Danish chemical production plant, Cheminova Ltd. due to its practices of discharge and emissions of toxic chemicals to the environment. This has been the case since its production started in the 1940's in a Copenhagen suburb, and up to the present day, where it is located in the North-western part of Denmark within an area which has been designated as a nature conservancy area of international importance under the Ramsur Convention.

In the following is a description of factory development and of problems associated with handling hazardous chemicals in an environmentally acceptable way. Emphasis will be placed on description of the individual sources and the focal points of pollution, and on the ecotoxicological impact and damages that has been caused by this pollution.

A summary of restrictions and precautionary measures which have been imposed on the factory in response to reported ecotoxicological accidents will be given also.

## DEVELOPMENT OF FACTORY AND OF ITS MAIN POLLUTION SOURCES

Early in its history, Cheminova Ltd. was donated to the University of Aarhus, Denmark. Formally, thus, it is owned by a foundation, and it is operated on a strictly commercial basis. It has grown to become a medium-size world producer of pesticide chemicals, specializing in the manufacture of thiophosphate insecti-

3

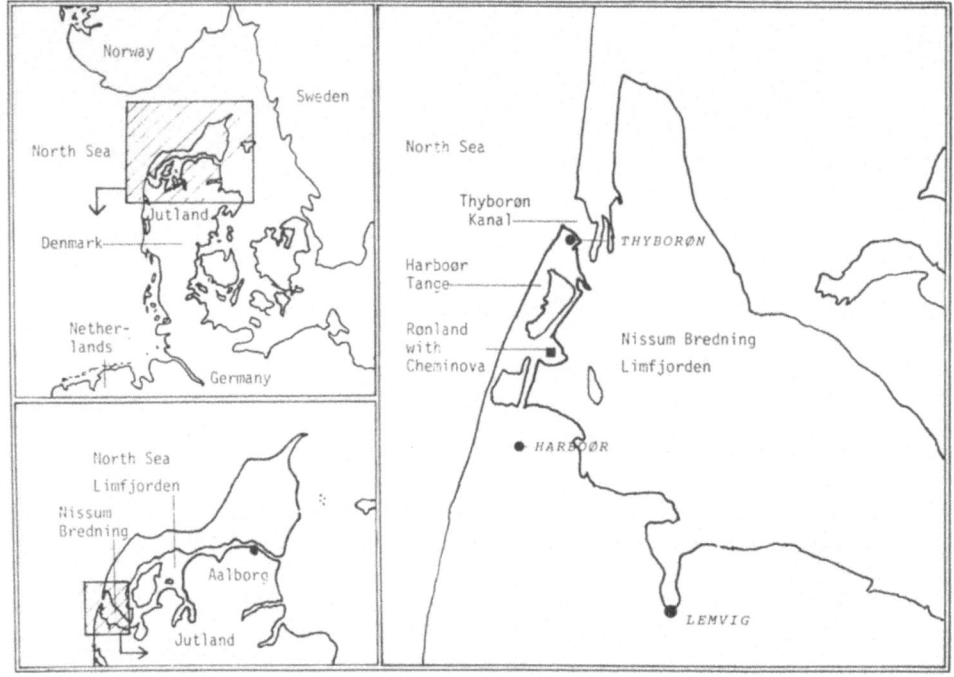

Fig. 1. Location of chemical plant Cheminova Ltd., Denmark

cides, espec. parathion, methyl-parathion, malathion etc. Early
productions of organomercurial fungicides, certain phenoxy acid
herbicides and others are to-day less significant, or they are
completely phased out.

In 1953 the factory moved from the Copenhagen region in dis-
grace with neighbours and local authorities, and it reestablished
itself on Harboør Tange in the Northern Jutland, *i.e.* on a strip
of land which is shaped as a sandy bar separating Limfjorden from
the North Sea (cfr. fig. 1). The factory was built close to the
new road connecting the fishing towns Harboør (a few miles South
of the factory) and Thyborøn (similarly towards the North). Both
the factory site and the road were built on land of filled-in sand,
and on the original bed of the fjord. The factory premises faces
the fjord, and the ground was stabilised along the coastline by a
concrete wall.

During the years 1960-62 this "old factory" was abandoned and
the production moved to a new site on the neighbouring peninsula
of Rønland, which stretches into the Nissum Broad, *i.e.* the shallow
waters forming the western part of Limfjorden. Today, all production facilities and factory premises are situated on Rønland
with the exception of the 2 - 3 km long pipeline, which leads the
waste water from the factory across Harboør Tange, and into the
North Sea.

In spite of its relatively remote position, and many precautions allegedly taken by the factory, the Cheminova development has been characterised by the continuous controversy with its neighbours and with authorities over hazards and threats to the environment and to local fisheries. It may be considered a culmination of this, that the Cheminova case in the years 1981-1983 was reexamined and scrutinized following an official order, which referred to a Government White Paper (March 1981) dealing with the history and the environmental practices of the factory (ref. 12).

The main problem areas have been the discharge of production waste water into the North Sea, they have been the leachings of toxic chemicals into the sea and into the fjord waters from various burial and dump sites, and they have been connected to the final disposition of solid waste depots. During the years, much concern has also been expressed on the penetrating and unpleasant smell, which is a characteristic of most thio-phosphorus chemicals. This latter aspect, however, will not be treated in this review.

An overall estimate on deposited amounts of chemicals and/or leaching rates in and around Cheminova is illustrated in fig. 2. Figures summarize the situation per 1981/82 on the basis of reports from different authorities.

## Factory waste water

Untreated waste water from Cheminova was discharged into the sandy dunes between two dikes on the North Sea coast from 1953 - 1962. During this period the factory established its production of organophosphate insecticides, and it initiated also (on an experimental basis?) productions of phenyl mercuriacetate, ethylmercuri-chloride, 2,4-D, MCPA etc., all of which contributed with chemical reaction- and by-products to the waste materials from the factory.

Based on a license issued by the authorities, the use of the locality between the dikes was extended, and during a 5-year period from 1957 - 62 it was further utilized as a dumping site for chemicals and toxic solid waste from the factory and from others. In 1962, the dump site - now called the "chemical depot at groyne nr. 42", or simply; the "groyne depot" - was closed by a sand cover, and a new discharge practice was initiated, *viz.* drainage of the waste water through a pipeline leading into the North Sea a few hundred meters outside the groynes. At the same time, a new dump site for disposal of solid waste, later known as the "Sulphur pits" (see below), was established on Rønland in the close vicinity of the factory buildings.

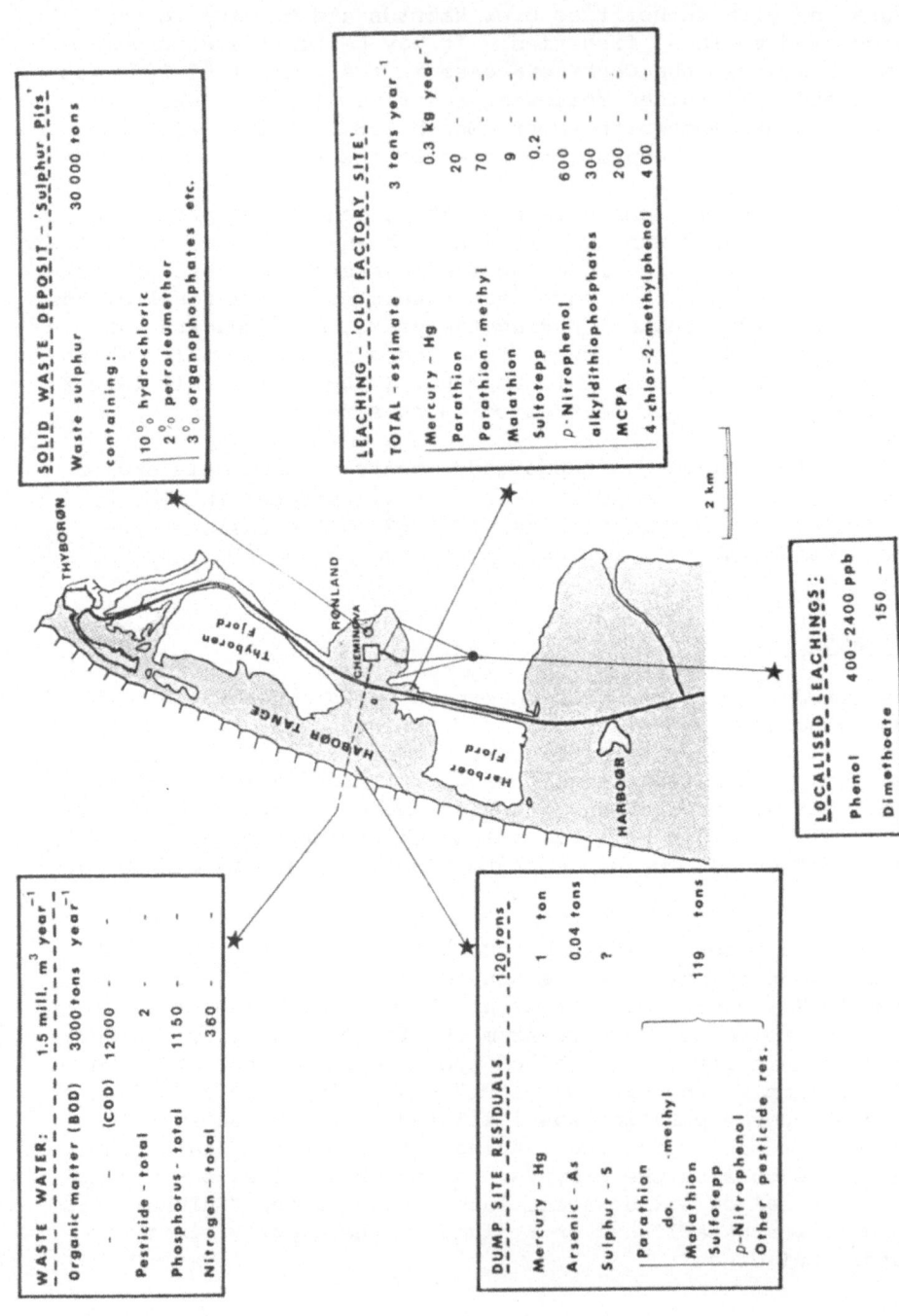

SOLID WASTE DEPOSIT - 'Sulphur Pits'

Waste sulphur          30 000 tons

containing:

10 % hydrochloric
 2 % petroleumether
 3 % organophosphates etc.

LEACHING - 'OLD FACTORY SITE'

| | TOTAL - estimate | 3 tons year$^{-1}$ |
|---|---|---|
| Mercury - Hg | 0.3 kg year$^{-1}$ | - " - |
| Parathion | 20 | - " - |
| Parathion - methyl | 70 | - " - |
| Malathion | 9 | - " - |
| Sulfotepp | 0.2 | - " - |
| p - Nitrophenol | 600 | - " - |
| alkyldithiophosphates | 300 | - " - |
| MCPA | 200 | - " - |
| 4 - chlor-2 - methylphenol | 400 | - " - |

WASTE WATER:          1.5 mill. m$^3$ year$^{-1}$

Organic matter (BOD)   3000 tons year$^{-1}$

- " -     (COD)    12000 - " -

Pesticide - total        2    - " -

Phosphorus - total     1150  - " -

Nitrogen - total        360  - " -

DUMP SITE RESIDUALS - - - 120 tons

Mercury - Hg            1 ton

Arsenic - As            0.04 tons

Sulphur - S             ?

Parathion - " - methyl
  do. - " - methyl
Malathion              119 tons
Sulfotepp
p-Nitrophenol
Other pesticide res.

LOCALISED LEACHINGS:

Phenol        400-2400 ppb

Dimethoate    150 - " -

2 km

THYBORØN

Thyborøn Fjord

RØNLAND

CHEMINOVA

HARBOØR TANGE

Harboør Fjord

HARBOØR

Fig. 2. Status on approximate amounts of deposited chemicals, rates of leachings and discharges into marine areas surrounding Harboør Tange (1981/82).

As far as the waste water was concerned, the change in discharge practices should soon prove to be of disastrous consequences. Already in the summer of 1963, complaints were reported on catches of fish with off-smell and off-tastes, and on dead or damaged lobsters along the North Sea coast around the plant, as well as from Limfjorden. In 1964 this was repeated with massive killings of fish and birds (cfr. table 1) reported from the vicinity of the pipeline outlet.

In 1965 the factory was brought for a local court, and it was established that residues of pesticide chemicals could be found in fish which were caught in wintermonths. The highly toxic Sulfotepp was found to be present in the waste water, whereafter Cheminova was forced by court decision to purify (and dilute) its waste water to a certain, specified toxicity level (reference: 1).

The officially accepted waste water discharges from the factory have been regulated by this court order since then. Recently, however, the wisdom of the court decision has been disputed again due to the findings that an average of 14% of all chemicals discharged or leached into the North Sea is carried by current and tide into the shallow and more vulnerable Nissum Broad.

Regular chemical analyses of waste water are only known from the latest year, *i.e.* since 1976. A summary of annual discharges of main chemical constituents is given in fig. 2.

## The first depot of toxic chemicals - or the "groyne 42 depot"

Before the dump site between the dikes at groyne nr. 42 was closed, a nationwide collection of old and left pesticide chemicals was initiated by the Ministry of Agriculture, and the collected chemicals were disposed of into the dump. By this, the Danish central authorities seem to have assumed a share of responsibility for the depot, although only in the order of 40 tons were collected, or corresponding to 3-4% of the total amount of chemicals later found in the depot.

The depot was attacked twice by winterstorms during the following years, and it was further opened incidentally in 1971 by the coast inspectorate. An asphalt cover imitating the shape of the surrounding dunes was made in 1971/72, but a continued leaching of chemicals from the depot and into the North Sea was not hindered. Full evidence of such leaching was presented in 1981 (cfr. fig. 3), and in the summer of 1981, the depot was finally opened following a government order, because it was considered a threat to the interests of the local fisheries. The massive part, or corresponding to about 80% of the contained chemicals was removed (see below).

Table 1.   REPORTED INCIDENTS OF BIRD AND FISH KILLS   (Reference 12)

| PERIOD | NISSUM BROAD | | NORTH SEA COAST | |
|---|---|---|---|---|
| | CATEGORY | REPORTED CASES | CATEGORY | REPORTED CASES |
| | Factory erected on "old site" in 1953/54.   Discharge of chemicals into coastal dunes | | | |
| 1954-1961 | Dead fish (plaice and flounder)<br>Dead lobsters<br>Dead eel<br>Off-taste fish | 1-3 x<br>several<br>1 x<br>regularly | | |
| | Factory moved to Rønland in 1960/62.   Waste water discharge into North Sea | | | |
| 1962-1965 | Off-taste fish<br>Dead fish<br>Damages to oyster banks | occasionally<br>3 x<br><br>1 x | Bird kills<br><br><br><br>Fish kills<br><br>Porpoise kills | 1962: 1 x  ( 36 birds)<br>1963: 2 x  (175   -  )<br>1964: 7 x  (100   -  )<br>1964: 1 x  (massive)<br>1965: 1 x  (  -  )<br>2 individuals |
| | Local court issues regulations on maximum toxicity of waste water, 1965.   "Sulphur pits" started | | | |
| 1965-1977<br>1978 | Occasional bird kills | | Few reports on off-taste fish<br>Extensive bird kills (gulls, terns, ducks)<br>Fish kills | <br><br>1 x |
| 1980 | Bird kills and damages to benthic fauna | | | |
| | Waste deposit in coastal dunes ("groyne depot") removed in 1981 | | | |
| 1981<br>1982 | Damages to benthic fauna<br>do. | | Occasional bird kills<br>Reports on mercury in ducks and seals | |

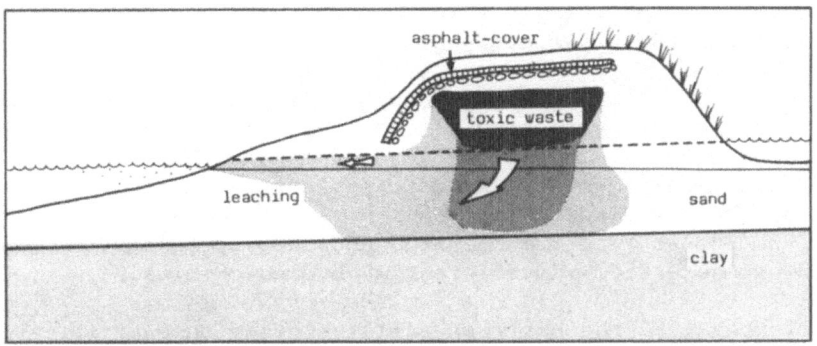

Fig. 3. Evidence for leaching of toxic chemicals, incl. mercury
from "groyne depot" in the dunes was found in 1981. Depot contained 500 tons of sulphur, 700 tons of organic
chemicals and 4 tons of mercury. (References: 4 and 22)

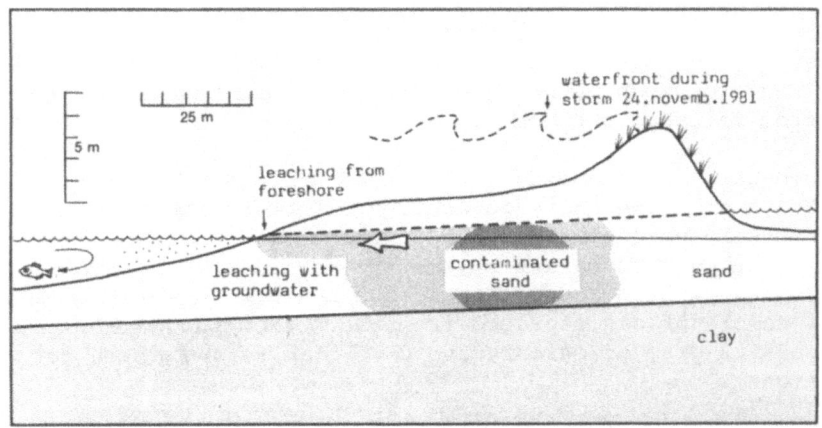

Fig. 4. "Chemical waste depot at groyne nr. 42" was removed in the
summer of 1981. Remaining deposits distributed in sand
and in foreshore are estimated to leach measurable in the
coming 10-40 years. (References: 4 and 22)

The remaining amounts of chemicals (or 20% of total) are now
left in the underground below the former depot or they are distributed under the foreshore. This is estimated to consist of about
1 ton of mercury chemicals, and about 120 tons of organophosphate
chemicals and metabolites etc. (cfr. fig. 4).

Chemicals deposited at the "old factory site"

As already mentioned, Cheminova operated its first productions
on Harboør Tange until 1962 on a lot of sand-filled land towards
Limfjorden. This site covers an area of 100 x 300 metres. In
1962, after moving to Rønland, the old production premises were

demolished and leveled with the ground. The area was covered with soil and left as "the old factory site".

During the following years, and until 1980 it was occasionally reported that a certain leaching of chemicals took place along the coast in front of the "old factory", and into the fjord waters. In 1967-68 a reinforcement of the concrete buttress was made, and the local authorities ordered the otherwise open area to be fenced to hinder the public from trespassing, and thereby to prevent incidents of exposure from insufficiently covered chemicals.

In November 1980 the buttress was stabilised once again, and sandy sediments and bottom material was scrabed from the fjord to strengthen the coast. This event disclosed that a heavy leaching and contamination of the fjord bottom had taken place. Through the massive disruption of the bottom, mercury chemicals, organophosphates, paranitrophenol and many other chemicals were released into the fjord ecosystem resulting in acute killings of crabs, worms, mussels, and other aquatic forms of life. The coastline sediments and the water, and especially the icecover of the fjord in the following winterperiods were all contaminated and heavily discoloured from released chemicals.

Discoloration was caused by the great amounts of $p$-nitrophenol (PNF), while the ecotoxic accidents to benthic fauna and flora more likely were caused by the presence of undegraded pesticide chemicals, such as parathion, methyl-parathion, malathion etc., all of which were found in sand and benthic materials. One individual sample of sand scrabed from the bottom showed a maximum value of 1.2 gram of mercury and 0.75 gram of parathion per kg (dry weight).

Chemicals from the underground of the "old factory" had obviously penetrated under the concrete wall and been brought to the surface of the fjord bottom by the groundwater movements under hydraulic pressure and influenced by the tide. In 1982, Cheminova resumed full responsibility for the chemical waste and the underground contamination, and measures against the rapidly developing situation was initiated.

## The second chemical waste depot - or the "Sulphur pits"

The Cheminova production of organic thiophosphates are chemically based on raw materials of phosphorus and sulphur (chemically reacted to form phosphor-pentasulphide or $PS_5$), various alcohols, chlorine, and other chemicals among which $p$-nitrophenol is the most important for the manufacture of parathion and methyl-parathion. In connection with the manufacture, a waste product is generated consisting of elementary sulphur (about 50%) contaminated with a number of organic phosphates and thiophosphates (3%), solvents (about 2%), and hydrochloric and acetic acids (about 10%), and water (about 35%) (cfr. fig. 2).

In a certain period after the factory moved from the "old factory" to Rønland, *viz.* from 1962 - 1966, the sulphur waste was deposited in natural or man-made hollows in the eastern part of Rønland, only a few hundred metres from the factory buildings and without any environmental control measures being taken. As from 1966, the waste was deposited in concrete tanks in order to carry out the local court order issued a year earlier. Plans were initiated with the intention to develop a process for extraction and reuse of the sulphur from the waste.

The concrete basins - the "sulphur pits" - have been used until recently, and it is estimated that a total of about 30.000 tons (other estimates state: 40.000 tons) of sulphur-rich chemical material have been deposited there during the 15-20 years period of active use (cfr. fig. 2).

During the years, a heavy corrosion of the concrete walls and of the bottom of the basins have occurred. In 1975 leachings of hydrochloric acid, organophosphates, and of sulphur were discovered into the underground surrounding the pits. Several attempts to revert the groundwater from flowing into Nissum Broad, Limfjorden have been made through protective pumpings and drain-systems, but leaching and heavy damages on the coastal fauna and flora have been seen up till the recent spring of 1983 (see below).

ECOLOGICAL DAMAGES AND ACCIDENTS

Complaints and reports on chemical emissions, of pollution, and of environmental damages in Limfjorden outside the factory exist already from 1954, *i.e.* the first year after establishment on Harboør Tange. With intermissions of relative calmness this has continued since then, and in 1962 the situation escalated to include damages on the North Sea side of Harboør Tange, as the immediate result of the above mentioned installation of the waste water pipeline.

Fish and bird kills

Among the first cases of fish kills and the numerous reports on off-taste fish, the most serious early impact of increasing pollution seem to have been the gradual disappearance of lobsters (*Homarus gammarus*) from the region. From a moderate, though steady commercial harvest of about 12-14 tons of lobsters per year in 1950-54, it dropped to practically zero in 1964. The species has never resettled in the area since that year.

This was the immediate cause for the authorities to institute

Table 2. PARATHION IN WASTE WATER CONTAMINATED FISH, 1964-65[o]

| Species | Oct. 1964 | Nov. 1964 | Dec. 1964 | Jan. 1965 |
|---|---|---|---|---|
| Cod (*Gadus morhua*) | 0.1[*] | | <0.005 | 0.06 |
| Liver from cod | 0.08 | | | |
| Plaice (*Pleuronectes platessa*) | 0.06 | | | 0.02 |
| Liver from plaice | 0.13 | | | |
| Flounder (*Platichthys flesus*) | | 0.02 | | 0.14 |
| Turbot (*Psetta maxima*) | | 0.005 | | |
| Dab (*Limanda limanda*) | | | <0.005 | 0.03 |

[o] All figures in mg kg$^{-1}$ (fresh weight) (Reference: 12)

[*] *para*-nitrophenol found in supplementary analyses are not included in reported values

investigations and to raise the court case mentioned above. Evidence was presented to the court that the Cheminova waste water required more than 50.000 - 100.000 x dilution to prevent acute toxic reactions to test species of fish, espec. guppies (*Lebistes reticulatus*). Parathion residues were found in many samples of commercial fish species caught outside Cheminova, e.g. cod (*Gadus morhus*), plaice (*Pleuronectes platessa*) and flounders (*Platichthys flesus*) (cfr. Table 2), but lobsters proved to be the most sensitive to the ecotoxicological influence of thiophosphates, espec. to the Sulfotepp impurities resulting from ethyl-parathion production processes.

Practically all incidents of fish kills, which were reported in the 1962-65 period were followed by reports of bird poisonings (cfr. Table 1) around groyne 42. Birds were most sea gulls and terns, but also a number of other fish-eating species, which are regularly nesting in the sanctuary formed by the marsh land between the dikes at the North Sea coast and the Rønland peninsula. Individuals of dead porpoises were also collected with signs of intoxication.

The court case seemed to be unique for that time in its decision to restrict and regulate Cheminova's waste water discharges on the basis of ecotoxicologically defined limits. Later experiences demonstrated the ability of the industry to meet the demands requested by the court, although serious cases did develop in the following years. These cases, however, have been explained by accidental production failures (*e.g.* in 1978, cfr. table 1), or they have been connected to situations of leaching from the different waste depots etc.

## Behavioural studies and residues in marine fauna

During the first years of Cheminova pollution around Harboør Tange, it was repeatedly observed by fishermen that lobsters behaved abnormal as early signs of intoxications. Symptoms were described as lack of balance, loss of claws after a few days, and death possiby within a week.

Such observations have attained renewed interest in recent years connected to the continued and chronic state of pollution. Changes in burrowing behaviour for benthic organisms, such as mussels (*e.g. Macoma baltica, Cerastoderma edule*, and *Abra alba*), and polychaetes (*Nereis diversicolor* and *Scoloplos armiger*), as well as significant increases in avoidance activity of fish (*Solea solea*) and shrimps (*Crangon crangon*) have regularly been noted in field studies (Reference: 15, 16). They have also been confirmed in laboratory trials as behavioural responses to several specific Cheminova chemicals of the thiophosphate type (cfr. figure 5).

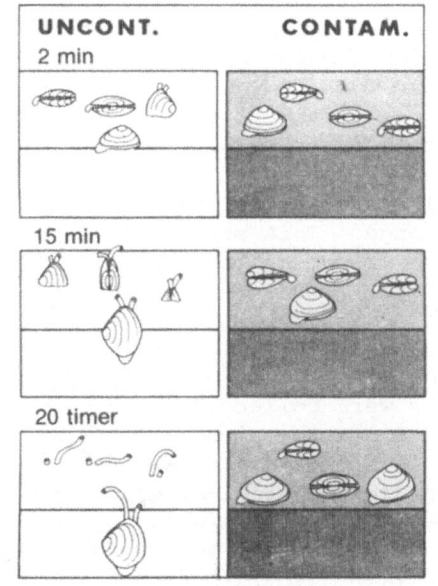

BURROWING RATE OF
*Macoma baltica*

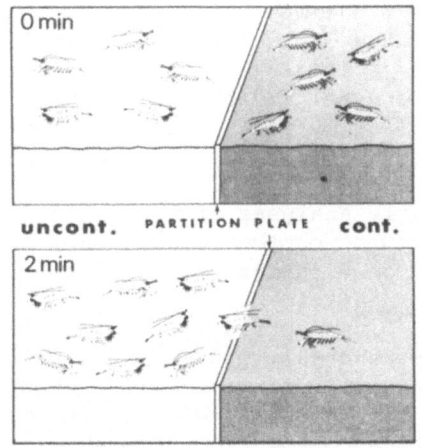

AVOIDANCE REACTION OF
*Crangon Crangon*

Fig. 5. BEHAVIOURAL RESPONSE TO PESTICIDE CONTAMINATION
(Reference: 15 and 16)

| NORTH SEA | 1 | 2 | 3 |
|---|---|---|---|
| Winter 1981 | 0.10 | 0.41 | 0.58 (0.02-2.0) |
| Summer 1981 | | | 0.44 |
| Summer 1982 | | 0.05 | 0.48 |
| Winter 1983 | 0.30 | 0.55 | 0.95 |

| NISSUM | B | C |
|---|---|---|
| Winter 1981 | 0.4 (0.2-0.8) | 0.8 (0.1-1.7) |
| Summer 1981 | | 0.7 |
| Summer 1982 | 0.4 | 0.3 |
| Winter 1983 | 0.4 | (0.2-1.1) |

Fig. 6. Mercury contents (in mg kg$^{-1}$ fresh weight) in flounders *(Platichtys flesus)* from coastal waters around Harboør Tange, 1981 - 1983 (Reference: 14).

By fishermen this has often been held to be significant as part of the explanations on disturbances of the ecological balance and the decreasing yields of commercial fisheries in recent years in the coastal region near to Harboør Tange. This situation has further aggravated since the observation of unacceptable chemical residues in fish and in mussels in 1980/81.

Early analyses of blue mussels *(Mytilus edulis)* during 1980 attracted attention because they indicated levels of definite con- tamination with mercury when the mussels were collected in the North Sea at groyne nr. 42, and - not the least - along the coast- line of Nissum Broad in front of the "old factory" (References: 6, 8, 9, and 23). The mercury was suspected to originate from de- posited waste and from buried chemicals, which after years in ob- livion had leached their way via the underground and penetrated the marine bottom from beneath.

In the following wintermonths of 1980/81 this suspicion was confirmed and reinforced by an extended analytical programme made by the authorities, and covering not only mussels, but also commer- cial fish species from the North Sea and from Limfjorden. Included

were further groundwater and subsoil samples from the depot sites
and foreshores.  Especially disturbing was the presentation of mercury levels up till 1.7 and 2.0 mg per kg of fish in the indicator
species flounder *(Platichtys flesus)*, which were caught around
groyne nr. 42 and in Nissum Broad outside the "old factory" site,
respectively (cfr. fig. 6).

Such levels exceed the hygienic threshold limit which are
administrated in Denmark for mercury-containing fish, namely 0.5
mg per kg.  The findings therefore provoked - by government intervention - that commercial fisheries were officially stopped in the
contaminated areas.  An immediate action was taken to stop further
leaching of chemicals around groyne nr. 42 by removal of the "groyne
depot", and thorough investigations were initiated to identify and
to quantify other suspected dump sites, espec. the "old factory"
and the "sulphur pits".

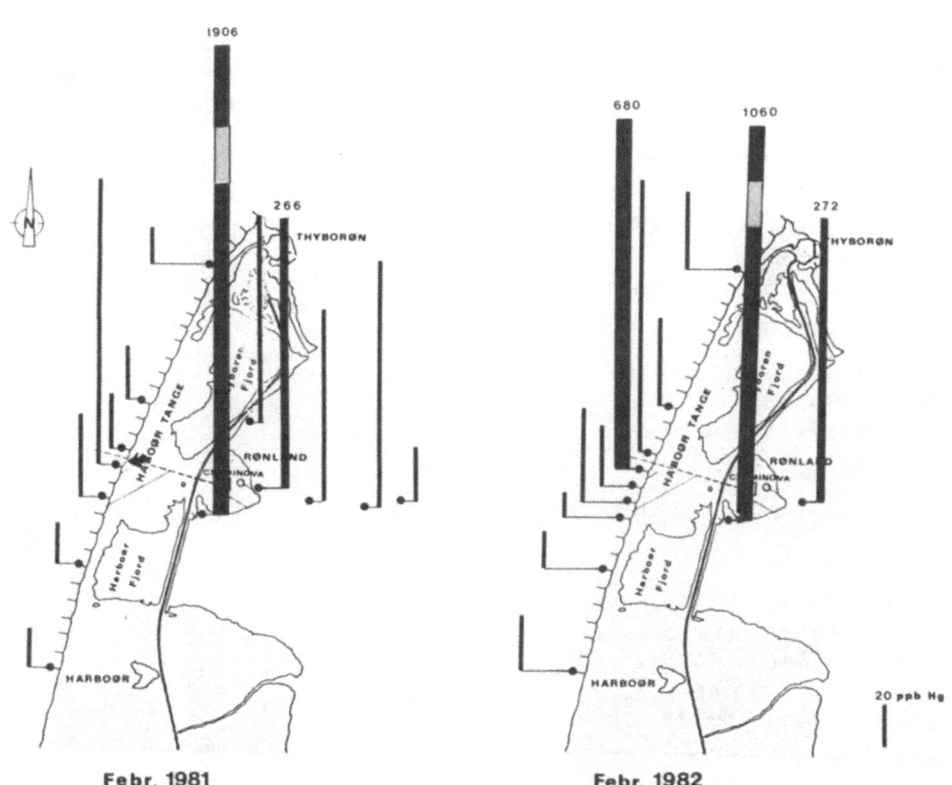

Fig. 7.   Mercury levels in mussels *(Mytilus edulis)* from North Sea
coast and from Nissum Broad - before and after the removal
of the "groyne depot" (at arrow)(Reference: 22 and 23).

It is interesting to note from the chemical analyses of *Myti-lus edulis* collected between the groynes (cfr. fig. 7) that the mercury contents in February 1982, *i.e.* 6-7 months *after* the removal of the depot had increased by a factor of 5-10 times compared to the contents the year before.  Obviously the opening and the removal of the depot had left a proportion (about 20%) of the originally deposited chemicals uncovered, and - which was not unexpected - more exposed to percolation and to leaching at a higher rate than before.

These observations were verified and found especially to be significant for wintersamples of mussels and flounders, probably caused by the higher rates of coastal water exchanges and the accellerated  groundwater movements under the influence of precipitation and a greater number of storms in the wintermonths.

Table 3. MERCURY IN MARINE BIOTA IN FRONT OF "OLD FACTORY", 1981-82

| SPECIES | No. of samples | Mercury mean, $\mu$g kg$^{-1}$ |
|---------|----------------|--------------------------------|
| **DEPOSIT FEEDERS** | | |
| *Macoma baltica* (Baltic sea mussel) | 2 | 1400 |
| *Abra alba* (bivalve) | 3 | 4400 |
| *Arenicola marina* (lugworm) | 2 | 760 |
| **SUSPENSION FEEDERS** | | |
| *Cerastoderma edule* (cockle) | 4 | 890 |
| *Mya arenia* (clam) | 5 | 940 |
| *Mytilus edulis* (mussel) | 3 | 1900 |
| **BROWSER** | | |
| *Littorina literea* (common winkle) | 1 | 2300 |
| **PREDATORS** | | |
| *Neohthys hombergi* (Polychaetes) | 4 | 770 |
| *Crangon crangon* (common shrimp) | 2 | 560 |
| *Platichthys flesus* (flounder) | 25 | 735 |
| *Anguilla anguilla* (eel) | 2 | 320 |
| *Zoarces viviparus* (vivip. blenny) | 1 | 590 |
| **MACROPHYTES** | | |
| *Zostera marina* (eelgrass) (values in dry weight) | ca. 50 | 100-1100 |

(References: 2, 9 and 23)

## DISRUPTION OF THE BENTHIC FAUNA

It was the increasing number of observations and the disturbing facts on growing pollution from the Cheminova chemical deposits
which in the years from 1978-82 accelerated investigations and precautionary measures to be taken.  Several of the signs were con-
spicuous, such as the discolorations mentioned above in sea water
and in icecovers on the fjord caused by the para-nitrophenol, or
the case of seabirdpoisoning in 1978 (due to an accidental discharge of parathion), or a partial uncovering of the "groyne depot"
by a winterstorm in 1980.  Other signs were of a more confirmative
nature, such as the analytical evidence presented concerning mercury and thiophosphate contaminations of marine organisms.

### The "old factory" coastline

Among the less advertised, although important events, were the
information to authorities in November 1980 concerning chemical
contamination and the disruption of benthic fauna in front of the
"old factory" site.  As already mentioned, this was caused by the
release of chemicals after scrabing of bottom materials and sediments which had been exposed to a continued leaching from the "old
factory".

From later analyses, it was estimated that a total rate of
leaching amounted to not less than 3 tons of toxic chemicals per
year from the "old factory" site (Reference: 7).  Studies were also
made to give a full description of the massive contamination of the
underground from which the leaching took place.  Elaborate chemical
studies, espec. for mercury in biota (cfr. table 3), and biomonitoring of the benthic fauna was initiated in the moribound sections
of the coastal ecosystem, and along the full coastline of Rønland
(cfr. fig. 8).

### Coastal zones around Rønland

It was an extension of these studies which revealed and described similar situations of disturbance of the coastal ecosystems which occurred in 1982 and in the winter of 1983.  In one
specific case, the unchanged insecticide chemical dimethoate was
found as a leaching product which drained together with cooling
water from the factory into the fjord  also  resulting in damages
to the benthic fauna.

In another case, a 3 ha great area was contaminated outside
Rønland, ca. one kilometre to the North of the "old factory".  It
was caused by heavy leachings through the concrete walls and through
the bottom of the "sulphur pits" during a number of years, and the
accident was triggered by an extreme-low tide situation (Reference:
19).  An increased upward hydraulic pressure of groundwater along

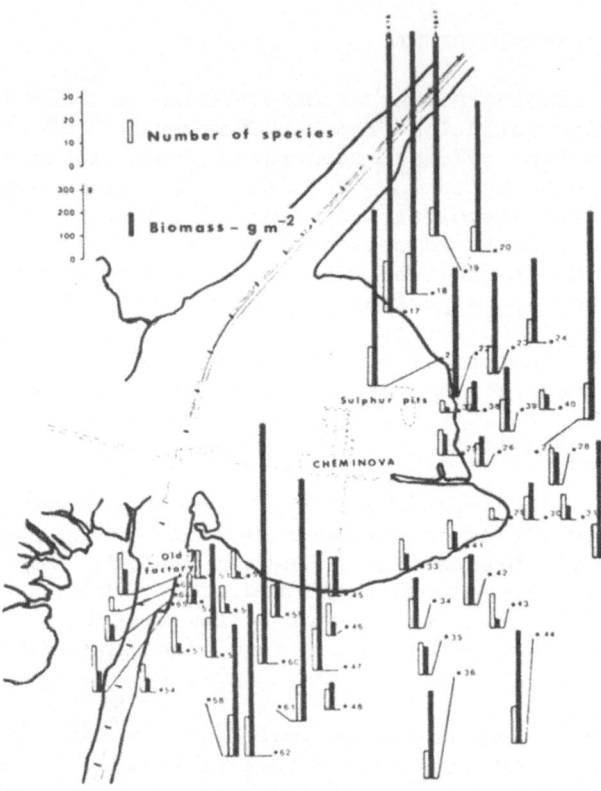

Fig. 8. Benthic Fauna - biomonitoring April/May 1981. Reductions in species number and biomass are indicative of deserted or dying localities outside "old factory" and "sulphur pits" Ref. 18 and 19).

the coastlines in Nissum Broad carried an eruption of toxic chemicals from the underground into the marine biosphere

The result of this was a complete or nearby complete killing of the benthic fauna in the coastal zone immediately to the East of the "sulphur pits". The damaged area was surrounded by severely inhibited zones with lowered biomass and reduced species diversity (Shannon-Wiener or Margaleff indices). Chemical analyses revealed significant contamination of groundwater and sediments with different esters of mono- and dithiophosphoric acids, *i.e.* hydrolytic breakdown products from parathion, malathion and similar insecticides.

The biomonitoring in such cases showed that mussels and certain worms and snails all were vulnerable and sensitive to the exposure of leached chemicals from the groundwater or sediments. The complete disappearance of such species as blue mussels *(Mytilus edulis),* cockles *(Cerastoderma edule),* and Baltic sea mussels *(Macoma baltica),* and the findings of dead or intoxicated crabs were reported as common observations.

In transitional zones which were moderately contaminated, inhibitions were measured with polychaetes *(e.g. Nereis diversicolor, Capitella capitata)* as the dominant among resistant species, including occassionally also cockles and Baltic sea mussels. Transitional zones could also be characterised by the presence of soft clams *(Mya arenaria)*and/or crabs which at finding were weak or damaged. Presumably situations have developed during which the animals have reacted to the advancing chemicals and have tried to escape.

PRECAUTIONS AGAINST FUTURE ACCIDENTS

Among variuos restrictions which have been imposed on the Cheminova factory during the years, those deriving from interventions of the National Danish Government in 1981 may be considered the most farreaching. The removal of the "groyne depot" was ordered, and a ministerial decree was issued that a complete review on production practices should be made in order to minimize and prevent any future accidents. Initiatives were finally taken to evaluate the risks of pollution from all deposited chemical waste and to take the necessary counter measures.

Since then, practical and technical precautions have been taken on that basis. These are mostly carried out in cooperation between the factory management and the Danish authorities. The most important among them are the precautions connected to the major dump sites, *viz.* the "groyne depot", the "old factory site", and the "sulphur pits".

Removal of the depot of groyne nr. 42

The removal of the groyne depot involved a major engineering effort. A total of 5.600 tons of sandy material containing 500 tons of elementary sulphur, about 700 tons of toxic, organic chemicals, and 5.4 tons of mercury and arsenic was removed from the dunes during the summer of 1981. The material was filled into metal drums for safe depositing.

As a criteriae which guided the removal, it was established that any remaining organic material in the sand should not exceed 1%, and following this decision, it was deemed unneccessary to remove material from subsoil below ground water level. About 120 tons of organic material was left as leaching chemicals in the sand below the depot and destributed into the foreshore.

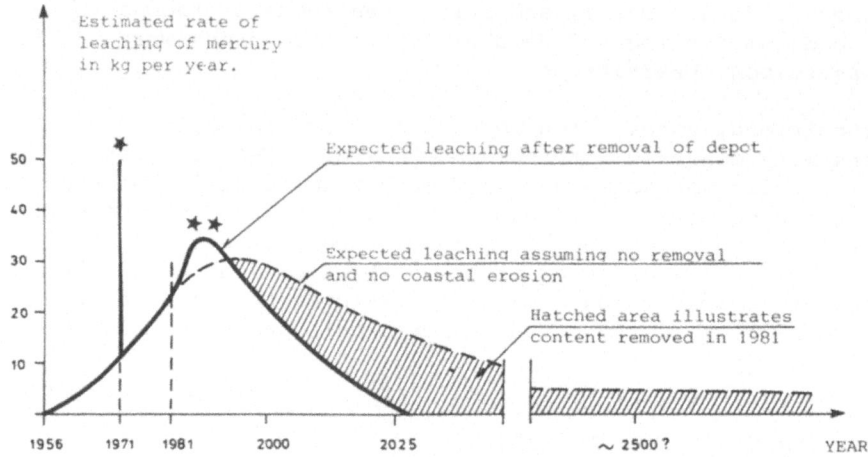

Fig. 9. Estimated mercury liberation to the North Sea from the
site of the "groyne depot". (Ref. 4)
  *) A certain part of depot was eroded into the sea in
     1971 whereafter depot was protected by an asphalt-
     cover.
 **) For a limited period after removal of depot in 1981,
     an increased rate of leaching is anticipated due to
     coastal erosion.

As indicated in figure 9 it was estimated that a certain leaching may continue from the foreshore into the North Sea for a period of approximately 40-50 years. On this basis, it has been pre-
dicted by calculation and by direct risk analysis that a release
of approx. 25 kg of residual mercury per year may be expected from
the former depot site and into the North Sea.

A similar analysis concerned with the rate of release has not
been made for the organic chemicals, although it seems likely that
there is a relatively higher risk of release for thiophosphates,
phenols etc. due to a higher mobility of these compounds in the
sand. This would especially be the case during the winterstorms
with increased coastal erosion and resulting in higher residual
contamination levels in the marine ecosystem. On the other hand,
the uncovering of the depot and the removal of the major part of
chemicals has most certainly reduced the time horizon for continued release, and it has diminished the total risk to a level which
is considered acceptable.

The costs incurred with the removal, inclusive transport and
final deposition in German salt mines of 25.000 drums totalled a
sum of about 23 mill. Danish kroner.

Table 4.    Estimated amounts (minimum values) of chemicals trans-
            ported with groundwater per year from old factory site
            into Nissum Broad, Limfjorden. (Ref. 7)

| | | |
|---|---:|---|
| Mercury | 0.3 | kg year$^{-1}$ |
| Parathion | 20 | kg year$^{-1}$ |
| Methyl-parathion | 70 | kg year$^{-1}$ |
| Malathion | 9 | kg year$^{-1}$ |
| Sulfotepp | 0.2 | kg year$^{-1}$ |
| para-nitrophenol | 600 | kg year$^{-1}$ |
| Total amounts of: | | |
| 0,0,0-triethylthiophosphate | | |
| 0,0-diethyl-S-methylthiophosphate | | |
| 0,0,S-triethylthiophosphate & | | |
| 0,0,S-triethyldithiophosphate | 300 | kg year$^{-1}$ |
| MCPA | 200 | kg year$^{-1}$ |
| 4-chlor-2-methylphenyl | 400 | kg year$^{-1}$ |
| Others-up to a total of: | 3.000 | kg year$^{-1}$ |

## Cleaning of the old factory site

In order to investigate the underground pollution of the "old
factory site", an extensive programme of groundwater sampling and
analyses was initiated in the summer of 1981. It was established
that the area was heavily polluted, and by hydrogeological methods
it was calculated that a minimum of 3 tons of organic chemicals
flowed with groundwater into the fjord per year. (Ref. 7).

The total amount of deposited chemicals was not established,
but about 90 compounds were identified (cfr. table 4). Chemicals
had penetrated and polluted the underground reservoirs of fresh-
water as well as of the sea water down to the impermeable layer
of clay at a depth of 12 meter below surface. The main part, however, of transport of chemicals into the fjord took place in the
fresh-water zone, i.e. in the upper 4-6 metres.

Release of chemicals from the "old factory site" is now considered to be prevented by means of a heavy and controlled pumping activity from a great number of borings in the area. Thorough
this it is achieved to control the movements of groundwater within the entire area. The rate of pumping is regulated according to
seasonal rainfalls, and it varies with the degree of pollution
around the individual borings.

Any organic material which is pumped from the ground is separated from the water-phase by sedimentation or centrifuging, while
the waterphase is purified by percolation thorugh activated carbon. The water is returned to the grounds or it is pumped and
mixed into the waste water pipeline leading from the present production facilities.

## Precautions and possible removal of the "sulphur pits"

The greatest amount of accumulated chemicals from the Cheminova productions is concentrated in the "sulphur pits". Since
1982 its use as an active depot has been discontinued, and the
management has initiated a practice of concentration and purification from the waste any surplus of elementary sulphur with the
purpose of reuse or re-sales.

The depot has now been covered and top-sealed in order to
prevent intrusion of rain or sea water into the massive part of
the depot. Simultaneously, a project of groundwater control has
been initiated in order to prevent penetration of water into the
depot from beneath and to protect against movements of polluted
groundwater in direction towards the fjord. As in the case of
the groundwater regulation on the "old factory site", all surplus
water is pumped away, purified and discharged together with factory waste water.

At present, there has been taken no final decision whether the
depot should be removed or not. The tendency is, however, to ad-
vocate for the complete removal of the depot, and gradually this
concept is accepted also by the factory management. This opinion
is based on present statements made by local authorities and not
the least by commercial fishing organisation who feel that the
risks connected to possible future eco-accidents will fully out-
weigh and overrule the costs which may be involved in a removal.

REFERENCES

1. Boetius, J. Toxicity of waste from a parathion industry at the
   Danish North Sea coast. Helgoländer wiss. Meeresunters. 17,
   182-187, 1968.

2. Brix, H. & Lyngby, J.E. Utilization of Eelgrass (Zostera marina) for monitoring of marine contamination with heavy metals
   (in the Danish). Urt. vol. 7, pag. 67-71, 1983 (Danish Bota-
   nical Society).

3. Bro-Rasmussen, Finn. Fisheries restrictions and maximum conta-
   minants limits in fish - an evaluation (in the Danish). Kaskelot. November 1982, pag. 16-23.

4. COWICONSULT Ltd. (Consultants for the Environment Protection
   Agency, miljøstyrelsen). Removal of Chemical Waste Depot at
   Groyne nr. 42 on Harboør Tange (in the Danish). Report of
   September 1981.

5. Degn, Erik. Problem areas in the Cheminova case (in the Danish).
   A review from Ringkøbing County Council, Environment office,
   DK-6950 Ringkøbing. May 1983. (In Dansk Natur - Dansk Skole,
   Annual report 1982).

6. Environment Protection Agency (miljøstyrelsen). Recipient description on coastal waters around Cheminova with special reference to chemical depot at groyne 42 on Harboør Tange. 1. and 2. report (in the Danish). April and November 1981.

7. Environment Protection Agency (miljøstyrelsen). Investigations of groundwater pollution on the "old factory site", Cheminova Ltd. Volume 1 and 2 (in the Danish). December 1981.

8. Kiørboe, Thomas. Mercury pollution around Cheminova (in the Danish). Kaskelot. November 1982, pag. 36-47.

9. Kiørboe, Th., Möhlenberg, Fl. & Riisgaard, H.U. Mercury levels in Fish, Invertebrates and Sediments in a recently recorded polluted area - Nissum Broad, Western Limfjord, Denmark. Marine Pollution Bulletin, 14, 21-24, 1983.

10. Isotopcentralen (IC/ATV). Distribution of Waste Water discharged from Cheminova Ltd. into the North Sea. Report nr. 8470 (in the Danish). May 1982.

11. Isotopcentralen (IC/ATV). Dissipation and dilution of Waste Water discharged from Cheminova Ltd. into the North Sea. Report nr. 151.70 (in the Danish). October 1982.

12. Ministry of the Environment. Government White Paper on the Cheminova case. Volumes 1, 2 and 3 (in the Danish). August 1981.

13. Ministry of the Environment. Memo to the minister concerning agricultural uses of contaminated areas on Harboør Tange (in the Danish). April 26. 1983.

14. Ministry of Fisheries. Fish for consumption from Coastal Waters around Harboør Tange (in the Danish). A report from an *ad hoc* Working group. August 1983.

15. Möhlenberg, Fl. Behavioural studies in Nissum Broad in front of the "old factory site", Cheminova Ltd. (in the Danish). Kaskelot, November 1982, pag. 26-35.

16. Möhlenberg, Fl. and Kiørboe, Thomas. Burrowing and Avoidance behaviour in marine organisms exposed to pesticide-contaminated sediment. Marine Pollution Bulletin, 14, 57-60, 1983.

17. National Chemical Supervision (Kemikaliekontrollen). Report on chemicals found in groundwater sampled on Rønland in August 1982. Report nr. 3-290L (in the Danish). April 7, 1983.

18. Ringkøbing County Council (amtsvandinspektøren). Monitoring of benthic fauna around Rønland in Nissum Broad (in the Danish).

19. Ringkøbing County Council (amtsvandinspektøren). Acute pollution in Nissum Broad around the peninsula Rønland. Report on lethal and sublethal effects on benthic fauna caused by leaching of chemicals (in the Danish). Report of March 1983.

20. Ringkøbing County Council (amtsvandinspektøren). Analyses of soil and grass around Nissum Broad and on Harboør Tange (in the Danish). Report of November 1982.

21. Riisgaard, Hans Ulrik. Biological effect studies in the marine environment. Strategies and methods in studies of sublethal effects on marine organisms. Thesis written in the Danish). Marine Pollution Laboratory, Copenhagen, 1982.

22. Riisgaard, H.U. The Cheminova pollution case (in the Danish). Kaskelot. November 1982, pag. 3-15.

23. Riisgaard, H.U. and Randløv, H. Mercury contamination of coastal waters around Harboør Tange and in Limfjorden measured by the monitoring of blue mussel (*Mytilus edulis*) (in the Danish). Report from the Marine Pollution Laboratory, Copenhagen. May 1982.

24. Water Quality Laboratory (VKI/ATV). Analyses of the Cheminova Chemical Waste Depot at groyne nr. 42 on Harboør Tange (in the Danish). Report nr. 81.540 of April 15, 1981. (In collaboration with the Analytical-chemical Services of Danish Civil Defense and the National Chemical Supervision).

# TETRAALKYL LEAD ACCIDENT IN SEA WATER

G. Tiravanti and R. Passino

Water Research Institute (IRSA)
National Research Council (CNR)
Rome, Italy

## INTRODUCTION

On July 14, 1974, the Jugoslav M:V "CAVTAT" (about 3,300 tons) collided at sun rise with the Panama freighter "LADY RITA" (about 11,000 tons) 5.6 km southeast of the Otranto Cape (Fig. 1) in the Adriatic Sea and sank at a depth of 94 m 5 hours after the collision. The cargo on deck and in the holds was 900 drums containing ~325 tons of tetramethyl lead (TML) and tetraethyl lead (TEL). The wreckage could have caused serious environmental pollution involving both the marine biota and the food chain.

Many background aspects of the problem, such as the pathway of circulation, degradation kinetics, residence time, final fate of Pb-alkyls (TAL) in the environment and long-term effects of TAL on the aquatic biota and human health, were completely unknown. It was impossible, moreover, to estimate the amount of TAL that ulimtately would be released into the environment following the sinking (Report of NATO Science Committee, 1976).

Using traditional analytical procedures (MIBK-APDC extraction and AA spectrophotometry), lead monitoring was performed on sea-water and on marine organisms 3 weeks after the accident (Tiravanti, 1974) and a few months later. No significant instantaneous TAL release was found (<10 uL). These results were not considered reassuring because specific analytical procedures for evaluating TAL concentration in sea water were lacking and because of very serious problems related to lead contaminations during collection of the sample, the storage, and the analytical operations (Tiravanti and Boari, 1979). Consequently, available analytical monitoring was considered inadequate for detecting pollutant release; specific analytical laboratory and field methods were strongly needed.

25

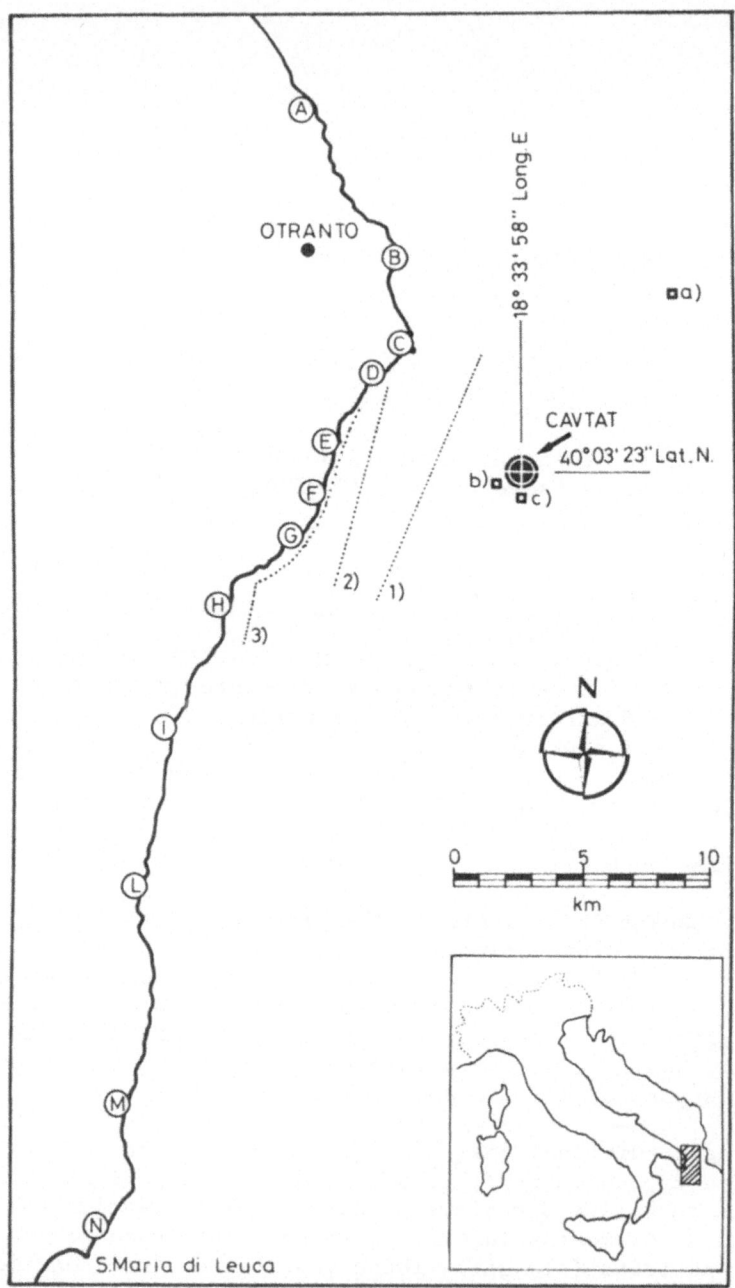

Fig.1. Map of the CAVTAT accident area with biological sampling locations. Key:  a), b), c)=cage locations; 1), 2), 3)= fish cruises; Ⓐ , Ⓑ , ..., Ⓝ = sampling locations along the coast.

Bioassays performed by Marchetti and colleagues (1978) indicated that TEL is much more toxic than TML. However, direct uptake of heavy metals by marine organisms from seawater is not the only route of contamination. Most pollutants are also absorbed from food or sediments (Bernhard and Zattera, 1975).

Patterson and Settle (1974) showed that only soluble lead interacts with biological activity and is accumulated by aquatic organisms. Assuming the same accumulation factors for the alkyls as for inorganic lead in phytoplankton, in the order of 100-1000 (Lowman and colleagues, 1971) and considering the TAL solubilities (Tiravanti and Boari, 1974), the higher trophic levels could be polluted by TAL in seawater, by their leaded food sources, or by products such as dimethyl mercury produced by interaction of TALs with the enviroment. Since no background data were available, studying a mathematical simulation of the marine and coastal pollution with highly conservative assumptions was the only available possibility for evaluating the phenomenon.

SIMULATION STUDIES

A mathematical model capable of simulating the diffusive and convective transport of released pollutant was developed (Tiravanti and Boari, 1976). The diffusion equation for a fluid, assuming constant fluid velocity u along the x axis and constant diffusion coefficients along the x, y, and z axes (Fig. 2) is:

$$\frac{\partial c}{\partial t} + u\,\frac{\partial c}{\partial x} = D_x\,\frac{\partial^2 c}{\partial x^2} + D_y\,\frac{\partial^2 c}{\partial y^2} + D_z\,\frac{\partial^2 c}{\partial z^2} \qquad (1)$$

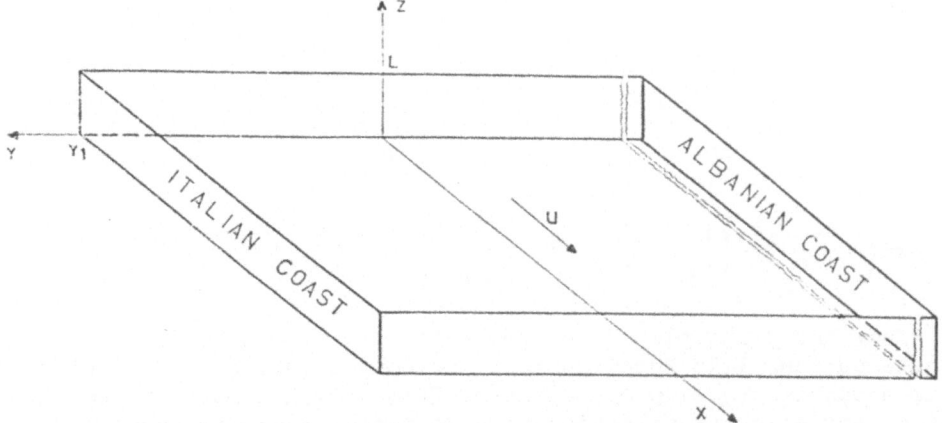

Fig. 2. Physical model of the Otranto Channel.

27

Because of the small depth variations in the wreckage area and because of the steep shore profile (Instituto Idrografico della Marina, 1962), the Otranto Channel was modelled as a rectangular parallelepiped L=93 m deep and 70 km wide (Fig. 2). The pollution source was located on the Channel bottom 5 km off the Italian coast. A steady state polluting flow was assumed. The stream velocity was assumed to be constant and parallel to the Channel walls as is the actual prevailing stream (leading southward) on the Italian side of the Otranto Channel (Mira-Zore Armanda, 1968). The channel sides, bottom, and water/air interface were assumed to be impermeable to the pollutant. These over-simplified assumptions gave overestimated concentrations in the channel system.

The following conditions were assumed from available data (Mira-Zore Armande, 1968; Talbot, 1974):
  stream velocity:  $u = 0.1$ m/s
  diffusion coefficient:  $D_x = D_y = 3$ m$^2$/s; $D_z = 0.015$ m$^2$/s

Equation (1) was solved both analytically and by numerical (finite difference) integration technique. In the analytical solution, the pollutant was assumed to be released as a series of pulses, and the effect of each pulse was summed to give the concentration due to a quasi-continuous release of pollutant. In the numerical integration technique, a finite continuous pollutant source was considered, and standard numerical techniques that solve Equation (1) for each node of the system were used.

## Analytical Solution

Equation (1) was first integrated considering the instantaneous emission of a finite mass of pollutant $M_P$ from a point source located at the coordinate origin. The concentration $C_M$ (x,y,z,t) for a single release at the point P (x,y,x) at the time t after the emission is:

$$
C_M(x,y,z,t) = \frac{M_P}{2\pi Lt(D_y D_z)^{\frac{1}{2}}} \left\{ \exp\left\{ -\frac{(x-ut)^2}{4D_x t} \right\} \right\} \left\{ \left\{ \exp\left\{ -\frac{y^2}{4D_y t} \right\} \right. \right.
$$

$$
\left. \left. + \exp\left\{ -\frac{(y+y_1)^2}{4D_y t} \right\} \right\} \right\} \cdot \left\{ \frac{1}{2} + \sum_{n=1}^{\infty} \left\{ \cos\frac{\pi n z}{L} \cdot \exp\left\{ (\frac{n\pi}{L})^2 D_z t \right\} \right\} \right\}
$$

Since the principle of superposition of effects holds for linear equations like Equation (2), the concentration distribution due to a continuous flow of pollutant from a point source is the limit conentration distribution of a series of instantaneous releases at the same mass flow rate when the emission frequency tends to ∞. The concentration distribution $C_G$ due to a series of N releases is:

$$C_G(x,y,z) \quad = \quad \sum_{i=1}^{N} \quad C_M(x,y,z) \qquad (3)$$

Because of computing time and related costs, a finite frequency
and a finite series of releases must be assumed. This assumption
leads to approximate solutions.


Numerical Solution

Equation (1) was first simplified by assuming steady state
conditions ($\partial c/\partial t = 0$) and negligible diffusion along the x axis
($\partial^2 c/\partial x^2 \ll u \partial c/\partial t$). For the numerical integration (finite difference
method) of the diffusion equation, the volume representing the
Otranto Channel was modelled as a grid, and spacing between the
node points was designed for desired density of node points at
various distances from the source for a minimum of computing time.
A finite (rectangular) pollution source, lying in a plane orthogonal
to the stream axis and centered at the coordinate origin, was used
in order to smooth concentration gradients (and consequently numerical errors) close to the source.

Figure 3 shows the pattern of computed TAL concentrations as
a function of distance downstream of the source in steady-state
conditions at 0 and 94 m depth, along the vertical plane comprising
the source and parallel to the stream direction. The continuous
line refers to the numerical solution; the dotted line refers to
the analytical solution.

In both cases, a safety factor as high as a release flow equal
to 100 kg/h of TAL in solution was assumed, corresponding to the
breakage of 1 drum every 90 min. This high flow rate was considered
to be very unlikely because of the mechanical properties and corrosion resistance of the drums and because of the low solubility of
TEL and TML in sea water (Grove, 1978). Figure 3 shows that concentrations computed with different assumed source characteristics
agree well about 1 km downflow of the wreck, where magnitude is
rather low -- about 10 ug Pb/L. It is interesting to note that,
while the lead concentration on the bottom is a monotonic decreasing
function, the surface concentration reaches a maximum approximately
at the point where complete merging concentration of the 94-m layer
occurs.

The three-dimensional graphs in Figures 4 and 5 show the
spatial concentration distribution on the sea bottom and on the
surface of the area downstream of the source, respectively, assuming a continuous release of alkyl lead (100 kg/h). On this diagram,
coastal concentrations are undetectable. Simulation tests were

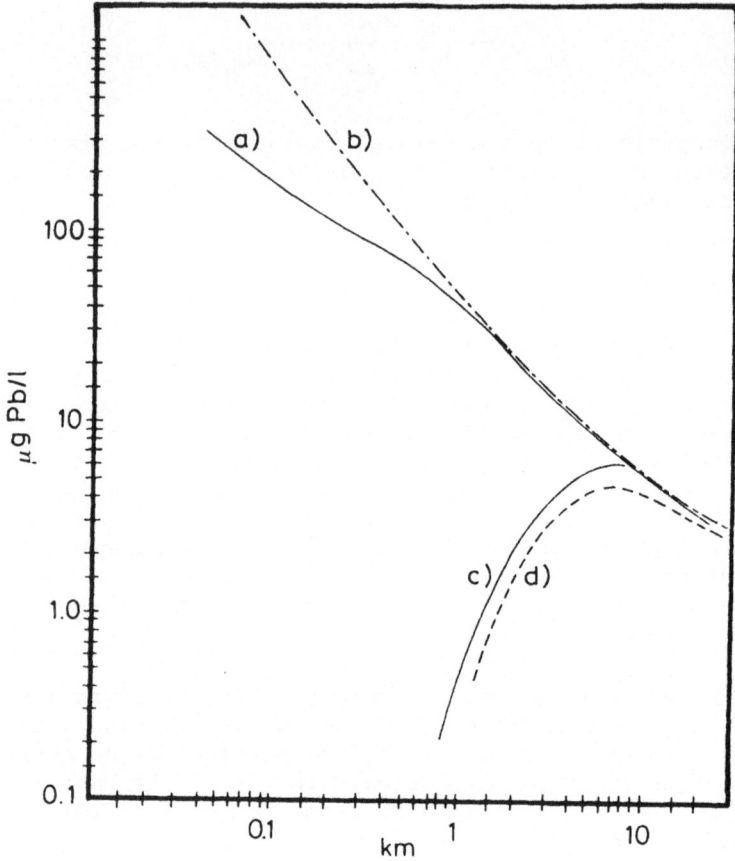

Fig. 3. Computed lead alkyl concentrations on sea water surface
(curves c, d) and along the sea bottom (curves a, b) as
a function of the source distance downflow. Lead release
flow: 100 kg/h. Continuous lines=numerical solution;
dotted lines=analytical solution. Reprinted with permission from Environmental Science and Technology, Vol.
13, 849-854 (1979). Copyright (1984) American Chemical
Society.

performed to determine the sensitivity of the model response along
the coastline to variations of the diffusion coefficients ratio
D/D and of the stream velocity u for the same G = 0.0278 kg/s high
flow rate. For the test case considered above (u = 0.1 m/s; $D_x$ =
$D_y$ = 3 m$^2$/s; $D_z/D_y$ = 0.005), a 4% difference between surface and
bottom concentrations was detected. For $D_z/D_y$ = 1, vertical mixing
was greatly enhanced, and bottom and surface concentrations were
coincident. Hence, the influence of the vertical diffusion coefficient on pollutant concentration distribution along the coastline

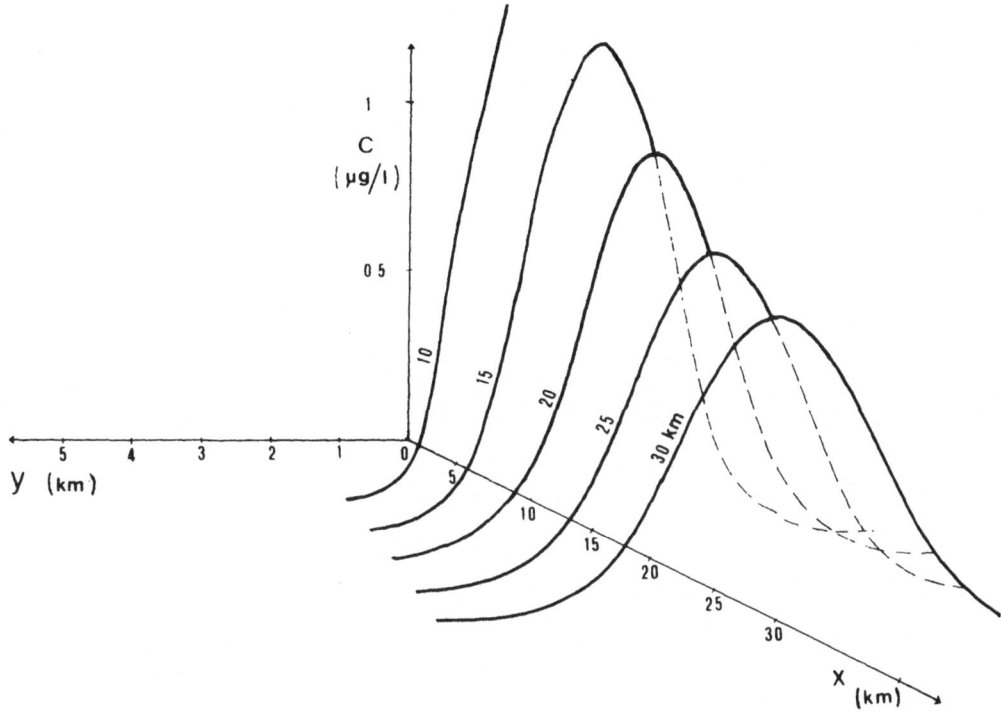

Fig. 4. Lead concentration distribution at the sea bottom within
the model of Otranto Channel. Data obtained from the
analytical solution of Equation (1) in the text. Direction
of the main stream = x; direction normal to the coast = y;
lead release flow = 100 kg/h. Reprinted with permission
from Progress in Water Technology, Vol. 12, 49-65 (1980).
Copyright (1984), Pergamon Press Ltd.

is practically negligible in the range $0.005 \leq D_z/D_y \leq 1$. This reduced
influence is mainly due to the shallow depth of the Channel compared
with its width and length. On the other hand, the effect of stream
velocity is considerable, as Figure 6 shows. In the test case
[stream velocity u = 0.1 m/s, which is comparable to the predominant
velocity in the wreckage area (Mira-Zore Armanda, 1968)], the
computed concentration along the coastline increased continuously
downstream.

The same qualitative trend was later observed on the alkyl lead
concentrations determined experimentally by Mor and Beccaria (1978)
on filtering living organisms collected along the coastline from

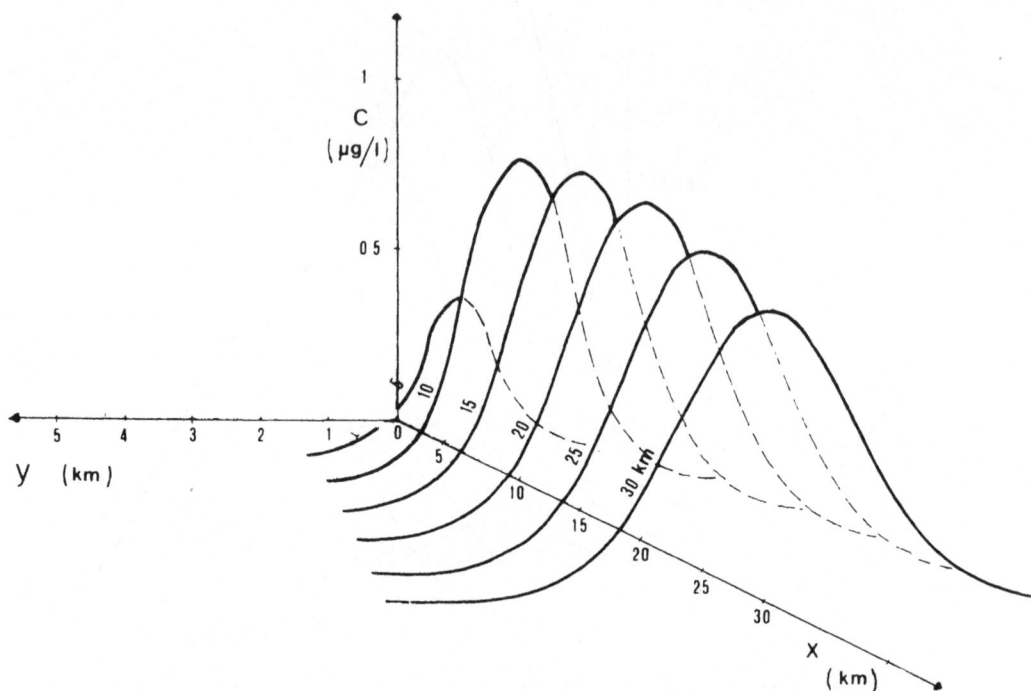

Fig. 5.  Lead concentration distribution at the sea surface within
the model of Otranto Channel.  Data obtained from the
analytical solution of Equation (1) in the text.  Direction
of the main stream = x; direction normal to the coast = y;
lead release flow = 100 kg/h.  Reprinted with permission
from Progress in Water Technology, Vol. 12, 49-65 (1980).
Copyright (1984), Pergamon Press Ltd.

Otranto to the Cape of Santa Maria di Leuca.  For lower velocities
(u = 0.05 m/s and 0.01 m/s), higher concentrations were computed,
but, even in the most unfavorable case, the maximum computed concentration along the coastline was <3 ug/L.

   Other experimental verifications of the model were not possible
due to:
   - unknown and undetectable alkyl lead releases
   - spatial and temporal variability of stream parameters
   - unknown diffusion coefficients
   - unknown interactions among alkyl lead-sea water-sediments-
     biota.  However, simulation tests indicated that coastal

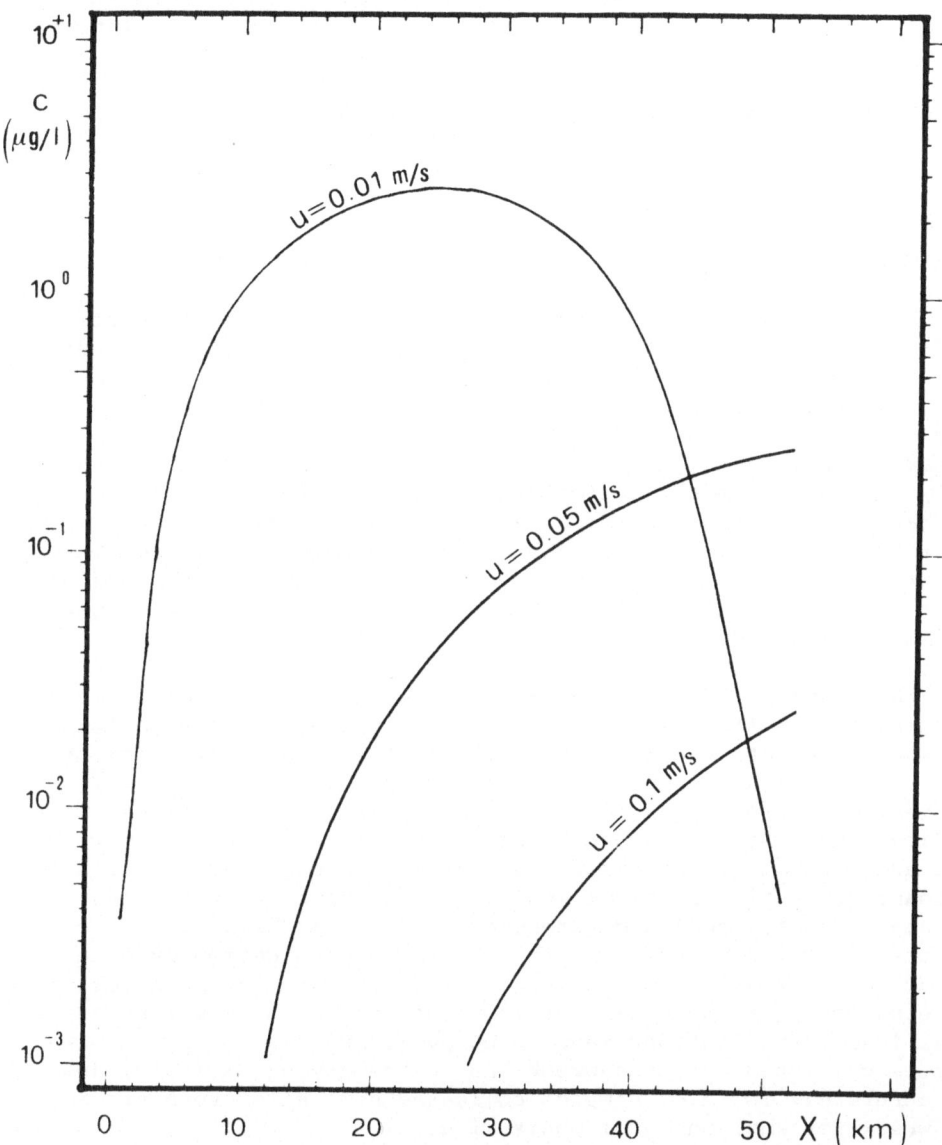

Fig. 6.  Lead concentration along the coastline, according to the
         analytical solution of Equation (1) in the text for
         different values of the stream velocity.  Reprinted with
         permission from Progress in Water Technology, Vol. 12,
         49-65 (1980). Copyright (1984), Pergamon Press Ltd.

pollution from alkyl lead release would not make bathing exposure
dangerous because of the very low concentrations even in conservative
conditions (high pollutant flow rate, high diffusion coefficients,
and low flow velocity). Nevertheless, direct toxic effects on the
marine environment and indirect effects on the coastal population
(particularly exposed groups like fishermen) through the food chain
could not be excluded. For this reason, rapid field and precise
analytical methods for organic and total lead determinations in
sea water were strongly needed.


CHEMICAL MONITORING

    Patterson and Settle (1976) demonstrated the extreme difficulty
in determining natural lead concentrations in both fresh and sea water
because of the wide possibilities of sample contamination by quantities of metal that easily change the order of magnitude of analytical
values. Contamination can occur during various operations: collection,
storage, handling (e.g., filtering), or addition of reagents and
analysis. Hence, most of the lead analyses performed by various
laboratories (Istituto di Ricerca sulle acque, Bari; Istituto
Superiore di Sanita , Roma; Laboratorio Centrale Idrobiologia, Roma)
before the recovery operations gave unreliable results because no
sampling and analytical techniques specific to lead in sea water were
available (Tiravanti and Boari, 1976).

    In October, 1975, sea water and sediment samples collected by
Yugoslav divers (Tomislav, 1975) close to the wreck were polluted
by lead (Tiravanti and Boari, 1976). This finding showed for the
first time that alkyl lead was released from the CAVTAT wreck.
Recovery operations (carried out later by SAIPEM Co., Milano) were
difficult because of the depth and the strong streams in the wreck
area and were dangerous because of the possible dispersion of significant quantities of alkyl lead at a low depth or at the sea
surface. Hence, local and extended pollution problems could be
expected: local concerns were for divers (contamination by skin
contact or by inhalation in the diving bell), and extended pollution
concerns were for the marine environment and the neighboring coast.
Alkyl lead determinations were to be performed: (a) rapidly in water
samples collected near the wreck and in the diving bell air, and
(b) accurately in water samples collected some kilometers around
the wreck for environmental monitoring.

    The necessity of carrying part of the analytical work on a ship
without sophisticated laboratory facilities gave rise to analytical
and organizational problems that had never occurred before. New
analytical methods were devised for solving some of the problems:
a rapid field test was designed for the analysis of TML and TEL in
sea water for safety purposes, and a complete analytical procedure,

reliable for Pb analysis in sea water accurate to tens of nanograms per liter, was developed for environmental studies (Brondi et al., 1981).

In the field test (Fig. 7), the sea water sample was filtered and oxidized in a separatory funnel. After elimination of the excess bromine, the pH value was adjusted to 9.3 with a buffer solution, and then dithizon in chloroform was added for the extraction-titration of the total heavy metals. In the lead colorimetric analysis with dithizon, the sensitivity is unfortunately much worse than for the zinc: 1 ug of zinc is equivalent to 3.7 ug of lead. For this reason, the results were expressed in terms of ug/L as zinc. Nevertheless,

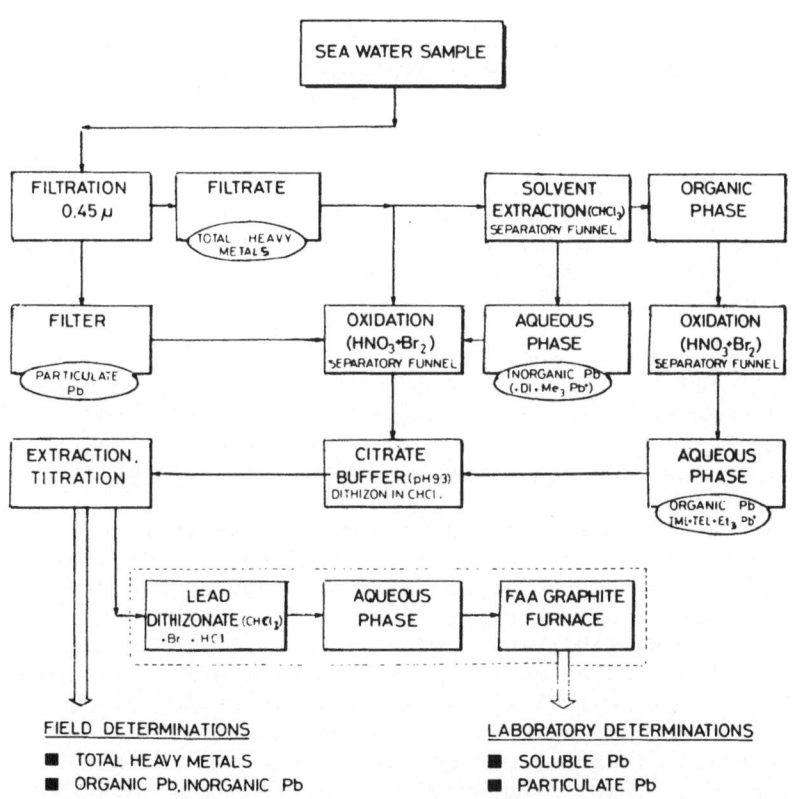

Fig. 7.  General scheme for field and laboratory analytical procedures.

this circumstance led to conservative conclusions for heavy metals analysis for protection of divers. The field organic lead determination was performed by a sea water extraction with chloroform and an oxidation of the organic phase, followed by a pH adjustment and by extraction-titration with dithizon in chloroform. In this way it was possible to determine TML+TEL+Triethyllead. The aqueous phase remaining after the extraction contained inorganic metals, trimethyl, and dialkyllead, which were determined in the usual way as an integration of the previous methods.

The above methods were checked with TML and TEL standard solutions, and recovery as high as 95% was found. The laboratory lead analyses were performed beginning with extracts of lead dithizonate in $CHCl_3$, following a re-extraction of lead in aqueous phase, and a flameless atomic absorption determination with a graphite furnace. This procedure minimized sample contamination and improved the sensitivity of the method.

Collection and preparation of many samples were carried out on the ships used in the recovery operations. Consequently, samples collected for analysis could have possibly been contaminated. It is worthwhile to note that laboratory blank values were in the order of 20-30 ng Pb/L, in spite of all the precautions taken in purifying the reagents and glassware.

Table 1 reports some field analytical results on sea water samples collected by divers around and inside the wreck. The average for heavy metals and for organic lead was 15 and 10 ug/L, expressed as zinc, respectively. These results precluded direct contamination of the divers.

Laboratory analyses showed the highest concentrations in samples collected in the holds were in the order of 5 ug Pb/L. For distances >500 m from the wreck, concentrations were always <0.1 ug Pb/L, and only the surface waters showed contents higher than this value, probably due to anthropogenic input and contamination.

Sediment analysis required development (Tiravanti et al., 1980) of light and manageable tools for sediment sampling close to the wreck because of the possibility of clogging the remote control samplers in the wreck structure. Total lead was determined on sediment samples with a mean value of 15 mg Pb/kg dry weight, which is normal according to sediment composition in this area (Tiravanti and Boari, 1979). However, some location values differed substantially from this mean value. Relatively high lead concentrations were also found in cotton bales loaded in the holds (Tiravanti et al., 1980) because of pollution by lead alkyls released from damaged drums.

36

Table 1. Some field analytical results on sea water samples collected by divers during the recovery operations. Analyses were carried out on lead alkyl extracted with chloroform (3), inorganic heavy metals in solution after extraction (2), and total heavy metals (1). Reprinted with permission from The Science of the Total Environment, Vol. 19, 21-31 (1981). Copyright (1984) Elsevier Science Publishers

| Date of Collection | Sample location on the wreck | (1) Total heavy metals (Cu+Zn+Pb) (ug Zn/L) | (2) Inorganic heavy metals (Cu+Zn+Pb) (ug Zn/L) | (3) TML+TEL+Et3Pb+ (ug Zn/L) |
|---|---|---|---|---|
| 27 Apr 1978 | Stern, leak side, near diving bell | 18.7 | 14.0 | 5.0 |
| 28 Apr 1978 | Deck, near drums | 20.0 | 10.0 | 8.5 |
| 28 Apr 1978 | Deck, near drums | 5.0 | 5.0 | 1.6 |
| 29 Apr 1978 | Deck, supernatant mud | n.d. | 150.0 | 214.0 |
| 29 Apr 1978 | Deck, near drums | 17.5 | 7.5 | 10.0 |
| 29 Apr 1978 | Near diving bell, after recovery operations | 10.0 | 8.7 | 1.3 |
| 7 Apr 1978 | Inside the containers with drums | 300.0 | 220.0 | 81.0 |

37

BIOLOGICAL MONITORING

Chemical-biological monitoring was carried out before, during, and after recovery operations. The goal of this biological monitoring was twofold:

(a) to assess any environmental lead alkyl population close to the wreck during recovery operations,

(b) to investigate the effects of recovery operations on the marine organisms collected along the Salento coast during and after recovery operations.

A new analytical method was then developed (Mor and Beccaria, 1978) to determine effects of lead alkyls on marine organisms. After a preliminary dehydration of the sample under vacuum, using a trap containing liquid $N_2$, the volatile lead was extracted with benzene, mineralized with 6 N HCl at high temperature, and then analyzed by polarography. The inorganic lead was determined on the dehydrated sample after digestion with the $HCl-HClO_4$ mixture, again using polarographic methods.

Soon after the start of the recovery operations, three cages (Fig. 8) with panels made with non-toxic asbestos concrete were immersed close to the wreck (11.3 km N; 0.8 km SW and S; see Fig.1) so that some substrate was available for growth of sentinel fouling organisms. These organisms should have been analyzed every two months after an initial six months period to allow growth of sufficient material, but, unfortunately, after eight months all equipment was lost in spite of every effort to find them. In any case, some data exist on the cages placed south of the wreck (Table 2). These data, coupled with data on organisms collected in the buoy, show that the fouling organisms near the wreck were polluted by lead. This indicates that some organic lead was lost during recovery operations, mainly on the sea bottom.

Table 2. Lead analysis on fouling organisms collected from the cages and buoys six months after the immersion, during recovery operations

|  | Station B | Station C | Buoy B |
|---|---|---|---|
| Location | 0.8 km | 0.8 km | 0.8 km |
| Organism | Chlamys | anomia | lepas anatifera |
| Organic Pb (mg/kg d.w.) | 44.8 | 55.9 | 10.5 |
| Inorganic Pb (mg/kg d.w.) | 144.9 | 109.7 | 23.9 |
| Total Pb (mg/kg d.w.) | 189.7 | 169.6 | 34.4 |

Fig. 8. Scheme of the cage: 1) floating buoy; 2) fiberglass
buoy; 3) sviwel; 4) cage with fin (V = 1 m$^3$; 5) anchor.


Some other data (Table 3) show that no pollution was observed
in pelagic organisms collected by divers near the hold of the
wreck (W. A. Flurst, Octel Co., London, private communication).

Table 3. Analysis of some organisms collected (W. A. Hurst, private communication, 1978) by divers near and into the hold of the CAVTAT wreck, during summer 1977

| Description of sample | Date of sample | ppm $^m$/m as Pb on wet basis | | | | |
|---|---|---|---|---|---|---|
| | | TEL | TML | $R_3Pb^+$ | $R_2Pb^{2+}$ | Total lead |
| Fish | 7.5.77 | <0.002 | 0.05 | 1.5 | 0.05 | 1.7 |
| Fish | 11.7.77 | <0.002 | <0.002 | 4.3 | 0.11 | 6.5 |
| Fish Q' | 16.5.77 | <0.002 | <0.002 | 0.03 | <0.01 | 0.01 |
| Fish 'R' (from well) | 23.5.77 | <0.002 | <0.002 | 0.02 | <0.01 | 0.09 |
| Fish 'S' | 12.10.77 | <0.002 | <0.002 | 0.02 | <0.01 | 0.04 |
| Fish 'T' | 11.7.77 | <0.002 | 0.005 | 0.03 | <0.01 | 0.05 |
| Shell fish from 'J' | 18.8.77 | 0.37 | 0.004 | 0.18 | <0.01 | 1.0 |
| Shell fish 'L' | -- | <0.002 | 0.004 | 0.34 | <0.01 | 0.74 |
| Shell fish 'M' (ex deck) | 14.9.77 | 0.006 | 0.006 | 0.49 | <0.01 | 0.56 |
| Shell fish 'P' | 7.5.77 | 0.02 | 0.009 | 0.33 | 0.03 | 0.66 |
| Star fish | 6.5.77 | 0.006 | 0.03 | 0.12 | 0.02 | 0.19 |
| Star fish | 13.5.77 | <0.002 | 0.05 | 0.22 | 0.02 | 0.42 |
| Sea urchin | 3.6.77 | 0.01 | 0.11 | 0.04 | 0.01 | 1.0 |
| Lobster | 6.7.77 | <0.002 | <0.002 | 0.57 | 0.01 | 0.86 |
| Ellesmere Port 'control'[a] | -- | <0.002 | 0.002 | <0.01 | <0.01 | 0.14 |
| Northwich 'control' | 16.5.78 | <0.002 | <0.002 | <0.01 | <0.01 | 0.02 |
| Northwich 'control' | 16.5.78 | <0.002 | <0.002 | <0.01 | <0.01 | 0.07 |

[a] Control = macherel.

Several collections of marine organisms along the coast were carried out during and after recovery operations (See Fig. 1). Edible fish and benthic invertebrates were caught by using a trawl and a grab. The analytical results (Panella, 1977) are presented in Table 4. The total lead content in the selected species (Callionymus maculatus, Mullus barbatus, Serranus hepatus, and Anellidi policheti)

Table 4.    Sampling cruise of 23-26 October 1978.   Total lead content
            in biological materials, expressed as mg/kg wet weight
            (mg/kg dry weight in brackets), collected near the
            CAVTAT wreck

| Species | Station A | Station B | Station C |
|---|---|---|---|
| Callionymus maculatus (length 64-85 mm) | 0.43 (1.77) | 1.63 (4.41) | 0.30 (1.13) |
| Mullus barbatus (length 142-174 mm) | 0.20 (0.83) | 0.16 (0.67) | 0.08 (0.32) |
| Serranus hepatus (length 55-74 mm) | 0.28 (1.20) | 0.16 (0.65) | <0.04 (0.15) |
| Anellidi policheti | 1.75 (10.0) | 1.09 (7.30) | 0.90 (5.60) |

was between 0.04 and 1.63 mg/L (wet weight).   These values are
normal for inorganic lead.   The lead alkyl concentrations determined
on Mytilus galloprovincialis and on Halocynthia pavillosa collected
along the Salento coast (Mor and Beccaria, 1978) were as high as
6 mg/L.   These concentrations were between 5 and 25 times higher
than those found on the same organisms collected in an unpolluted
area, near Argentario, Toscana.   It is worth noting, however, that
some organic lead was also found in the "blank" organisms.   This
means that the alkyl pollution of marine biota is a quite general
phenomenon, ascribed to either the use of lead alkyls mixed in
gasoline or to some natural methylation of organic lead in the
environment (Wong et al., 1975; Thompson and Crerar, 1980; Jarvie
and Whitmore, 1981).

    Table 5 reports some statistical data of the recovery operations
and shows that 7% of the products were missing.   The total cost of
the recovery was about 10 million US dollars (in 1978).


CONCLUSIONS

    The CAVTAT accident was a case of unpollution of the environ-
ment since ~93% of the 325 tons of the TAL were recovered.   A high
percentage of the 22 tons not recovered was lost immediately after
the accident, while low quantities not easy to recover were burnt
in the air.

    Before recovery operations began, potential marine and coastal
pollution were evaluated by a mathematical model.   Simulation tests
showed that even under unfavorable conditions alkyl lead concentra-

Table 5.  Statistical data of the CAVTAT recovery operations

| | TML-CB | TEL-CB | CR50-CB | Total |
|---|---|---|---|---|
| Loaded drums | 496 | 274 | 130 | 900 |
| TAL weight (tons) | 178.064 | 100.284 | 47.060 | 325.408 |
| Recovered drums | 496 | 270 | 107 | 873 |
| Empty drums | 18 | 11 | 3 | 32 |
| TAL recovery (tons) | 171.811 | 97.310 | 34.364 | 303.485 |
| TAL recovery % | 96.5 | 97.0 | 73.0 | 93.3 |
| TAL lost (tons) | 6.253 | 2.974 | 12.696 | 21.923 |

tions along the coastline would be <3 ug/L; hence, direct toxic effects on coastal population could be excluded, but interfering effects on trophic levels of the marine ecosystems had to be evaluated.  The model was solved both by analytical and numerical (finite difference) methods.  The analytical solution required considerably less software development and computing time than the numerical one.  However, the analytical solution made use of a more homogenous model with constant velocity and constant diffusion coefficients.  The finite difference method is better fitted to more complicated models (with variable velocities and diffusion coefficients).  Finite difference methods can be used when reliable experimental data are available and a more accurate approximation is required.

Mathematical models can be used to evaluate potential pollution hazards when:
(a)  a model is a simplified but sufficiently realistic representation of the system under investigation,
(b)  concentration levels of pollutants to be determined in the system are beyond the ranges of available analytical (chemical) techniques,
(c)  biological and/or physico-chemical accumulation of the pollutants are expected but cannot be detected and monitored in short time periods.

Specific field and laboratory procedures were developed for lead analysis in sea water.  Data from several surveys showed that alkyl lead pollution occurred only very close to the wreck and mainly during the salvage operations.  Indeed, total lead measurements in sea water performed from a few hundred meters up to some kilometers from the wreck always showed lead levels < 0.1 ug/L; only surface waters had lead levels of ~0.1 ug/L, probably due to anthropogenic inputs.

Pollution from the CAVTAT wreck was in the same order as that observed in marine environments near highly populated coasts. Average lead concentration values in sediments collected in the wreck area were 15 ppm, which is a normal value for marine sediments. Higher values, as expected, were found in sediment samples collected within the wreck ship.

Investigations carried out on edible fauna before and after the recovery operations gave normal lead concentration values that excluded significant pollution in the marine environment. Simulation results from the mathematical model and experimental data could not be directly compared, but the former showed that coastal pollution was very unlikely while the latter showed that significant pollution in the wreckage area was to be excluded before, during, and after recovery operations. This incident, although causing no harm to both man and the environment, points out the risks involved with toxic chemicals in sea transportaiton. Should these incidents recur, the methodological approach follwed in this case should prove valuable.


ACKNOWLEDGEMENTS

A special word of thanks to the personnel responsible for analytical and monitoring operations of the following Institutions: Water Research Institute (IRSA) of the National Research Council, Bari; Istituto Superiore di Sanita^, Rome; Laboratorio Centrale di Idrobiologia, Rome; Comitato Nazionale Energia Nucleare, Rome.


REFERENCES

Bernhard, M., and Zattera, A., 1975, in: "Marine Pollution and Marine Waste Disposal," Pearson and Frangipane, eds., Pergamon Press, New York.

Brondi, M., Dall'Aglio, M., Ghiara, E., Mignuzzi, C., and Tiravanti, G., 1981, Sci. Total Environ., 19:21-31.

Grove, J. R., 1978, Investigation into the formation and properties of aqueous solutions of lead alkyls, in: "Proceedings of an International Experts Discussion on Lead; Occurrence, Fate and Pollution in the Marine Environment," M. Branica, ed., Pergamon Press, Oxford.

Istituto Idrografico della Marina, 1962, "Carta IIM, 2 Ed.," Foglio 28, Genova.

Jarvie, A. W., and Whitmore, A. P., 1981, Environ. Technol. Lett., 2:197-204.

Lowman, F. G., Rice, T. R., and Richards, F. A., 1971, "Radioactivity in the Marine Environment," National Academy Press, Washington, D. C.

Marchetti, R., Chiaudani, G., Da Gasso, R., De Paolis, A., Gaggino, G. F., Gerletti, M., Provini, A., and Vighi, M., 1978, Mar. Pollut. Bull., 9:206.

Mira-Zore Armanda, 1968, Food and Agricolture Organization of the United Nations, Rome, VII.

Mor, E. D., and Beccaria, A. M., 1978, A dehydration method to avoid loss of trace elements in biological samples, in: "Proceedings of an International Experts Discussion on Lead; Occurrence, Fate and Pollution in the Marine Environment," M. Branica, ed., Pergamon Press, Oxford.

Panella, S., 1977, Report on the research performed in the Otranto Channel in the CAVTAT wreckage area, Rome (in Italian).

Patterson, C., and Settle, D., 1974, Contribution of lead via aerosol deposition to the southern California bight, N. 2426, Publication of the Division of Geological Planetary Sciences, C.I.T., Pasadena, Calif.

Patterson, C., and Settle, D., 1976, Special publication #422, 321, National Bureau of Standards, Washington, D. C.

Report of the NATO Science Committee's Panel on Marine Sciences on the Recommended Strategies in the CAVTAT Episode, 1976, Venice, Aug. 30-31.

Talbot, J. W., 1974, Paper #32 presented at the International Symposium on Discharge of Sewage from Sea Outfall, London.

Thompson, J. A. J., and Crerar, J. A., 1980, Mar. Pollut. Bull., 11:251-253.

Tiravanti, G., 1974, IRSA Internal Report (in Italian).

Tiravanti, G., and Boari, G., 1976, Judicial Report deposited to Otranto Magistrate's Court, Oct. 14 (in Italian).

Tiravanti, G., and Boari, G., 1979. Environ. Sci. Technol., 13(7): 849.

Tiravanti, G., Rozzi, A., Dall'Aglio, M., Delaney, W., and Dadone, A., 1980, Prog. Water Technol., 12(1):49-65.

Tomislav, K., 1975, Report on the CAVTAT wreck submarine survey, Brodospas Co., Split, Yugoslavia, Brindisi (in Italian).

Wong, P. T. S., Chau, Y. K., and Luxon, P. L., 1975, Nature, 253: 263-264.

## Nomenclature

$c$      concentration, $ML^{-3}$

$C_G$      concentration due to continuous release source, $ML^{-3}$

$C_M$      concentration due to instant release, $ML^{-3}$

$D_i$      diffusion coefficients across the plane normal to the i direction (i=x,y,z), $L^2 T^{-1}$

$G_P$      mass flow rate of pollutant, $MT^{-1}$

$L$      depth of the model channel, $L$

$M_P$      mass of pollutant released by an instant pulse, $M$

$t$      time elapsed after an instant release, $T$

$u$      stream velocity along the x axis, $LT^{-1}$

$x$      length coordinate of the channel, $L$

$y$      width coordinate of the channel, $L$

$y_1$      distance from the coastline to the pollutant source, $L$

$z$      height coordinate of the channel, $L$

# ENVIRONMENTAL CONTAMINATION BY TCDD: THE SEVESO CASE

Flaminio Cattabeni

Inst. of Pharmacology and Pharmacognosy, University of Milan, Via A. Del Sarto 21, 20129 Milano, Italy

## INTRODUCTION

Over the past thirty years there have been about 20 accidents in chemical plants manufacturing chlorinated phenols (1). The main toxic compound responsible for the health effects is a by-product, 2,3,7,8-tetrachlorodibenzo-p-dioxin (TCDD). TCDD is always formed during alkaline hydrolisis of tetrachlorobenzene to produce sodium trichlorophenate (TCP), according to the following scheme:

1,2,3,4,5 - tetra-chlorobenzene → (Ethyleneglycol, Na OH) → Na - 2,4,5 - tri-chlorophenate → (Hydrolysis, HCL) → 2,4,5 - tri-chlorophenol

Na - 2,4,5 - tri-chlorophenate → (>180°C) → TCDD

The amount of TCDD formed as a by-product is dependent upon the temperature at which the reaction takes place. In case of temperature above 200°C, the dimerization reaction of TCP to form TCDD preceeds at a high rate and with the production of heat. Under these circumstances, the amount of TCDD formed can become several points per cent.

TCDD is by far the most toxic compound synthesized by man. In animal experiments, the dose that kills 50% of the treated animals ($LD_{50}$), is in the order of $\mu g/kg$. In guinea pig, the most susceptible species, the $LD_{50}$ is around 0.5 – 1 $\mu g/kg$, whereas in dogs it is about 1000 $\mu g/kg$. This points out one of the many unique features of TCDD toxicity: the effects vary greatly according to the species considered (2). The sensitivity of the human being is not known: it seems however to be intermediate between the two extremes cited above.

Another remarkable feature of TCDD is its long persistency in the organism, due to its high lipophilicity and the absence of reactive groups on the molecule. So, metabolism and excretion are very slow and this gives rise to bioaccumulation in case of chronic exposure, even at low levels.

Despite the intensive investigations on the mechanism responsible for the lethal effects of TCDD, there is no definitive explanation for this (see Ref. 3 for review). Even after a single and high dose of TCDD, death occurs from two weeks to two months thereafter: none of the toxic effects (decreased body weight, induction of microsomal enzymes, tumor-promoting activity, thymus atrophy, effects on plasma membranes, reduced immunocompetency) is responsible per se for the lethal effect.

The health risks associated with TCDD in TCP production were known since 1957 not only by toxicologists, but also by managers of the chemical companies involved and safety measures were suggested and adopted by many TCP producers (1). Nevertheless, in 1976 one of the most dramatic ecoaccidents originated from a chemical plant producing TCP.

## CASE HISTORY

ICMESA (Industrie Chimiche Meda Società Azionaria) was a small chemical company located in the municipality of Meda, a densely populated village about 20 km north of Milano (Fig. 1). ICMESA is owned by Givaudan, a large

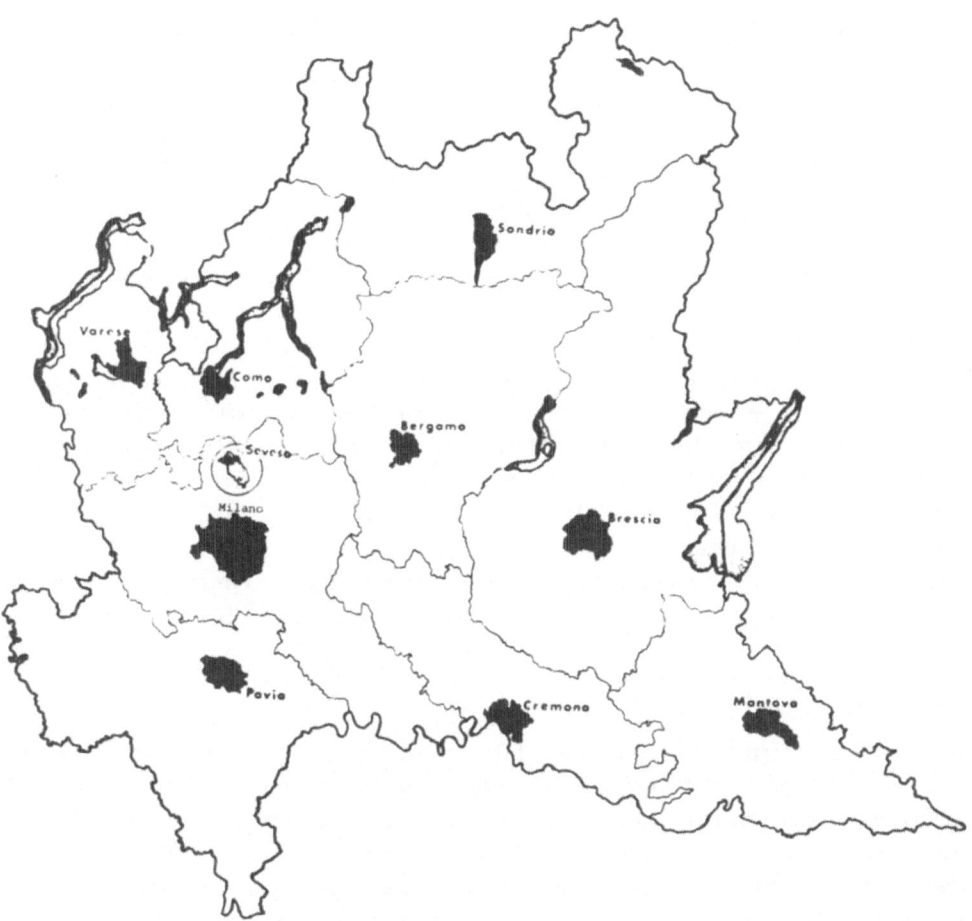

Fig. 1. The Lombardy Region: the area of Seveso (in circle) is just above Milano, the Region's capital.

swiss chemical company and a subsidiary of Hoffmann-La Roche. Hoffmann-La Roche and Givaudan managers decided, at the beginning of the seventies, that the ICMESA plant should start to manifacture TCP. The product was to be shipped to the USA and utilized exclusively for the production of hexachlorophene, an antimicrobial agent for topical use. The hydrolysis of tetrachlorobenzene to sodium-TCP was performed in a 10 ton reaction vessel. The TCDD content was carefully monitored at the research laboratories of Givaudan in Switzerland on every batch of TCP produced by ICMESA. Therefore, there is no doubt that swiss managers both at Givaudan and ICMESA were fully aware of the potential risks in the manifacture of TCP (4). Despite this, a safety valve connected to the reactor, which was designed to vent when the internal pressure exceeds 3.8 kg/cm sq., was venting straight into the air through the roof of the building.

The cycle of TCP production was shut off for the week-end at about 6 a.m. of saturday July 10. The alkaline hydrolysis was ended and about 15% of the solvent (ethylene-glycol) was distilled off. Therefore, the reaction vessel contained trichlorophenate, sodium chloride, excess sodium hydroxyde and ethylene-glycol. For an as yet unknown reason, the temperature in the mass rose above 200°C, the pressure within the vessel exceeded the safety limit and the rupture disk of the valve broke at 12:40 p.m. of the same day.

A mixture of vapors and fine liquid particles escaped through the orifice at sonic velocity. The plume of material reached a height of about 50 m above the roof of the building. This sonic emission was followed by an emission of vapors well below sonic velocity and in this case fairly large particles and droplets were emitted (5). Emission lasted for about two hours. A breeze blowing to south – south-east at about 10 km/hr dispersed the vapors and particles over a densely populated area belonging to 11 different municipalities; however, the major proportion of TCDD was found in the municipality of Seveso.

There are no definitive figures on the total amount of TCDD formed and relased in the environment. From a comparative evaluation of the analytical data on soil and vegetation in the area polluted by TCDD and the content of the reaction vessel after the accident, it has been estimated that the amount of TCDD released varies between 0.3 to 3 kg. Of this amount, about 0.150 kg has been deposited in the surroundings of the ICMESA factory (4).

## MEASURES TAKEN AFTER THE ACCIDENT

For a week after the release of the "toxic cloud" - as it was called - nobody was informed by either ICMESA or Givaudan about the possibility that together with TCP and sodium hydroxyde also TCDD could have been released. The first warning came on July 16 from Dr. G. Reggiani, the Medical Director of Hoffmann-La Roche, who informed physicians taking care of 11 hospitalized children that the skin lesions observed in their patients could be due to contact to an unknown quantity of TCDD (1). At this time, the first environmental samples were already analyzed at Givaudan Research Headquarters in Switzerland. However it was only on July 23 that Givaudan admitted the presence of high concentrations of TCDD in the environment and suggested that local authorities evacuate the population and inform the pregnant women of the possibility of an embryotoxic effect. These measures were not considered necessary by local authorities (1).
On the other hand, the presence of toxic materials in the environment was clearly indicated by deaths among domestic animals and some wildlife present in the area. Indeed the local population raised poultry and rabbits for houshold consumption. Three days after the accident several rabbits fed with vegetation collected around the factory died with pathological syndromes never observed before (6). Animal mortality became an important biological indicator of the extension of the contamination produced by the accident and it served as a basis for the chemical analysis of TCDD in the environment.
A first map of TCDD contamination was provided by Givaudan on July 23. Additional analytical work was also started on July 22 in two italian laboratories: Laboratorio Provinciale di Igiene e Profilassi (LPIP, Director: Dr. A. Cavallaro) and the Institute of Pharmacology and Pharmacognosy (Director: Prof. R. Paoletti), School of Pharmacy, University of Milano. Indeed it was only at that time (two weeks after the accident) that both health officers and local authorities begun to recognize that the accident resulted in one of the most serious environmental contaminations.

It became evident that thousands of samples were to be analyzed in a short time, in order to restrict the heavily contaminated area from human exposure. The analytical technique chosen for these analyses was mass-fragmentography, since the combination of gas-chromatography with mass-spectrometry provided the best requisites of both sensitivity and specificity. Environmental samples were collected, extracted with dichloromethane, the solvent was evaporated and the samples were reconstituted with a small volume of the same solvent (7). With this simple and rapid procedure, about one thousand samples were analyzed in two weeks. Work was carried out twenty-four hours a day in eight hour shifts. The approximate analytical sensitivity at that time was on the order of 0.1-0.5 ppb. This sensitivity was sufficient to give an estimate of the extension of the contamination for TCDD levels in areas at high risk. The highest level found was 50 ppm.

The analytical work provided the basis for local authorities - with the help of toxicologists - to divide the area into three different zones (Fig. 2):

1. Zone A: contamination greater than 50 μg/m sq.

2. Zone B: contamination between 50 and 5 μg/m sq.

3. Zone R: contamination below 5 μg/m sq.

As mentioned before, due to the large extent of the area that was possibly contaminated by TCDD, environmental sampling was performed in places where deaths of animals were reported.

After this emergency phase, a more systematic approach to determine both the extent and the severity of the contamination was started. Therefore, the sampling procedure was standardized (7) and the analytical procedure was refined (8), reaching a sensitivity on the order of 1 ppt. The analytical task force was composed of people from the above institutions and those from the Milan section of the Istituto Superiore di Sanità-Rome and from the M. Negri Institute for Pharmacological Research. On the basis of the data so collected, TCDD contamination was defined and a population of 736 persons was eventually evacuated from zone A (87.3 hectares), 90% belonged to the municipality of Seveso. The population (4699 inhabitants) of zone B (269.4 hectares) was

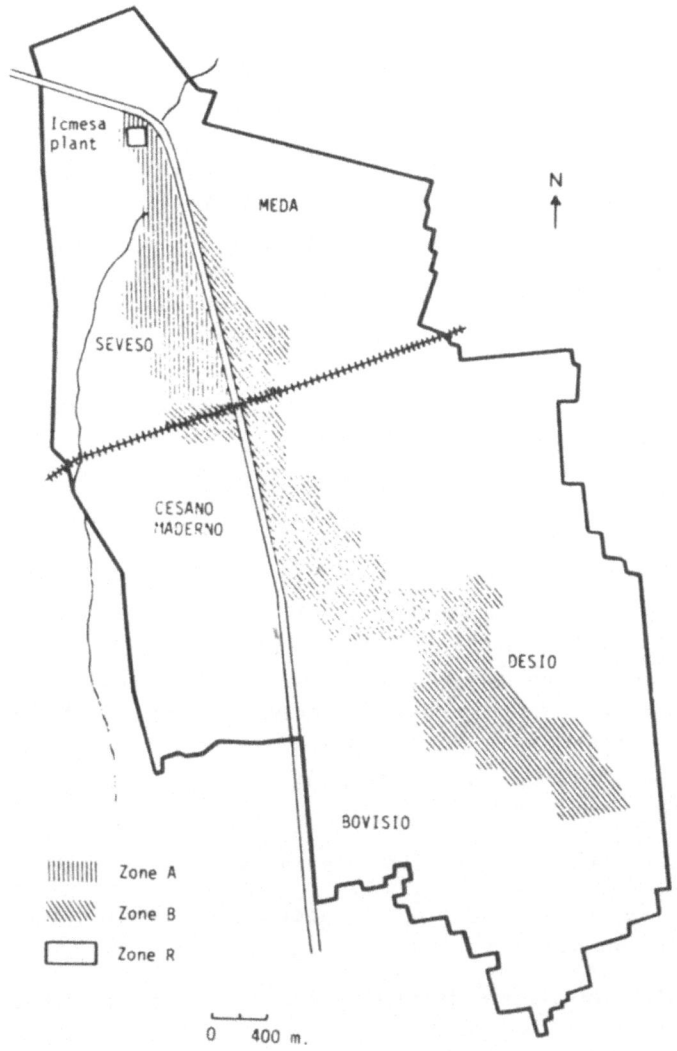

Fig. 2. Map of the area contaminated by TCDD and its sub-
division in zones A, B and R.

not evacuated but they were restricted from eating vegetables grown in their gardens and farm animals raised in their court-yards. Children aged between 1 and 12 years were removed from the polluted area during day-light hours in order to avoid their contact with the contaminated soil during open-air playing. Women were not allowed to live in zone B during the first 3 months of pregnancy. All animals were slaughtered (about 30,000). The area adjacent to zones A and B was defined as zone R. This area covered 1430 hectares with a population of 31,800 inhabitants. They were also prohibited from eating their own vegetables and meat.

Taking into account the population indirectly affected by the accident living in the 11 municipalities, about 220,000 people have been involved. A medical commission was tasked with the health surveillance of this population.

All the activities related to the Seveso accident have been and still are coordinated by a Special Office of the Regione Lombardia and located in Seveso.

## ACTUAL SITUATION

About 50 hectares of zone A are still (october 1983) highly contaminated. A yellow plastic fence still delimits this area. The other part of zone A has been recovered, removing the top soil: 511 persons moved back to their homes. The heaviest contaminated areas of zone B have also been decontaminated by the removal of top soil. All the contaminated material is now stored in zone A.

Those issues most debated today are:

1. How long will TCDD persist in the area?

2. Is it possible to establish an acceptable value for TCDD in the environment?

3. What are the long-term health effects due to the exposure to TCDD?

With regard to the first question, it seems that six months after the accident, a steady state of TCDD persistence in the soil has been reached (9). So, substantial amounts of TCDD may persist for ever in the area. Indeed, no microbiological degradation seems to occur for TCDD (10). The answer to the second question is of great importance not only for the residents in the Seveso area, but also for those who live in other areas were high levels of TCDD have been reported (Viet-Nam, several places in the USA and Canada). A Regional Committee, established ad hoc, has set, as a first step towards this decision, that an acceptable limit for daily intake of TCDD is 1 pg/kg/day. This figure has been obtained by dividing by an arbitrary safety factor of 1,000 the dose of 1 ng/kg/day, which appears to be a no effect level for daily consumption of TCDD in laboratory animals.

The decision about the level of TCDD in the environment compatible with human activities in the absence of health risks is a more arduous one. The main difficulty stems from the calculation of how much of the TCDD present in the environment is available for human intake, i.e. the transfer factor from the environment to man. Related to this problem is the long-term hazard due to TCDD. The crucial point in this regard is the carcinogenicity of TCDD, which has been demonstrated, but only at relatively high doses. Moreover, there is no explanation yet of the possible mechanism of cancer induction by TCDD (11). Without these elements, the long-term risks associated with the presence of TCDD in the environment remain puzzling.

On the other hand, acute toxic effects suffered by the population exposed at the time of the accident seem to be largely confined to chloracne. Of the 193 persons with chloracne attributed to TCDD exposure, about 20 are currently presenting active symptoms. Studies on congenital anomalies and mortality did not show increases above background levels (12).

These results can be interpreted either as indicating that human sensitivity to TCDD is much lower than that of experimental animals, or that the population was not exposed to sufficiently high doses of TCDD to cause adverse effects other than chloracne, which seems to be the most sensitive indicator of exposure.

## CONCLUSIONS

The case described here is the only accident where considerable amounts of TCDD have been released in the environment. Although the toxicity of TCDD was a known fact among scientists at that time, the lack of cooperation ties created much confusion in the first two weeks after the accident. The main lesson to be learned from this episode is that knowledge about toxic hazards from industrial processes has to be improved, with the improvement of the toxicological awareness in the general public.

## REFERENCES

1. Holmstedt, B., 1980, Prolegomena to Seveso, Arch. Toxicol., 44:211-230

2. Greig, J.B., 1979, The toxicology of 2,3,7,8-TCDD and its structural analogues, Ann. Occup. Hyg. 22:411-420

3. Matsumura, F., 1983, Biochemical aspects of action nisms of 2,3,7,8-tetrachlorodibenzo-p-dioxin (TCDD) and related chemicals in animals, Pharmac. Ther. 19: 195-209

4. Homberger, E., Reggiani, G., Sambeth, J. and Wipf, H.K., 1979, The Seveso accident: its nature, extent and consequences, Ann. Occup. Hyg. 22:327-370

5. The ICMESA accident (Seveso). Final report prepared for the Lombardy Regional Council on the accident, its consequences and recommended remedial actions, Cremer and Warner Ltd, 140, Buckingham Palace Road, London, October 1976.

6. Fanelli, R., Bertoni, M.P., Castelli, M.G., Chiabrando, C., Martelli, G.P., Noseda, A., Garattini S., Binaghi C., Marazza V. and Pezza F., 1980, 2,3,7,8-Tetrachlorodibenzo-p-dioxin: toxic effects and tissue levels in animals from the contaminated area of Seveso, Italy, Arch. Environ. Contam. Toxicol. 9:569-577

7. Adamoli, P., Angeli, E., Bandi, G., Bertolotti, A., Bianchi, E., Boniforti, L., Camoni, I., Cattabeni, F;, Colli, G., Colombo, M., Corradi, C., De Angelis, L., De Felice, G., Di Domenico, A., Di Muccio, A., Elli, G., Fanelli, R., Fittipaldi, M., Frigerio, A., Galli, G., Grassi, P., Gualdi, R., Invernizzi, G., Jemma, A., Luciani, L., Manara, L., Marinelli, A., Merli, F., Nicosia, S., Rizzello, F., Rossi, C., Rossi, G., Salvatore, G., Sampaolo, A., Schmidt, G.P., Taggi, F., Tebaldi, G., Zaino, E. and Zapponi, G., 1978, Analysis of 2,3,7,8-tetrachlorodibenzo-p-dioxin in the Seveso area, Ecol. Bull. 27:31-38

8. Di Domenico, A., Merli, F., Boniforti, L., Camoni, I., Di Muccio, A., Taggi, F., Vergori, L., Colli, G., Elli,

G., Gorni, A., Grassi, P., Invernizzi, G., Jemma, A., Luciani, L., Cattabeni, F., De Angelis, L., Galli, G., Chiabrando, C. and Fanelli, R., 1979, Analytical techniques for 2,3,7,8-tetrachlorodibenzo-p-dioxin detection in environmental samples after the industrial accident at Seveso, Analyt. Chem. 51:735-740

9. Di Domenico, A., Viviano, G. and Zapponi, G., 1982, Environmental persistance of 2,3,7,8-TCDD at Seveso, In "Chlorinated Dioxins and related compounds. Impact on the environment", Hutzinger et al. Eds. pp. 47-54, Pergamon Press, Oxford.

10. Hutter, R. and Philippi, M., 1982, Studies on microbial metabolism of TCDD under laboratory conditions, In "Chlorinated Dioxins and related compounds. Impact on the environment", Hutzinger et al. Eds. pp. 87-94, Pergamon Press, Oxford.

11. Fox, J.L., 1983, Dioxins health effects remain puzzling, Science 221:1161-1162

12. International Steering Committee for the Seveso accident: Reccomendations issued at its 6th meeting held in Milan, February 19-21, 1984.

MASS BIRD MORTALITIES ON THE MERSEY ESTUARY UK -

INCIDENT, INVESTIGATION, RESOLUTION AND FOLLOW-UP

Daniel Osborn

Nerc, Institute of Terrestrial Ecology

Monks Wood Experimental Station
Huntingdon, CAMBS PE17 2LS, UK

## ABSTRACT

1. In autumn 1979, a large mortality incident involving wading birds, wildfowl and gulls occurred on the Mersey estuary in north-west England. Smaller mortalities occurred in subsequent years. Dead and sick birds contained elevated levels of lead. The lead was mostly in the form of alkyl lead.

2. Apparently healthy birds from the estuary also contained elevated lead levels and showed signs of suffering from some of the internal abnormalities that dead and sick birds had exhibited. This raised the possibility that several thousand birds were at risk from alkyl lead pollution.

3. Experimental work showed that the alkyl lead in the birds had caused the mortalities and, further, it was concluded that, if further mortalities were to be avoided, lead levels in bird tissues should not exceed 0.5 mg/kg (wet wt).

4. Follow-up monitoring studies suggest that elevated lead levels and evidence of adverse effects on the birds were still being found some time after the amount of alkyl lead effluent had been reduced.

# INTRODUCTION

Many of Britain's off-shore islands and coastal cliffs are the
sites of internationally important seabird colonies. Similarly,
British estuaries form a network of internationally important
wintering grounds for wading birds, and many British estuaries
also support large populatons of wintering wildfowl and tens of
thousands of gulls. The location of the major wader and seabird
concentrations are shown in Figure 1.

Occasionally, large numbers of the birds are washed ashore dead
and dying. Sometimes it is obvious that they have been killed by
oil, but in some cases cause is far less obvious. There are 3
main reasons why the latter type of incident gives rise to
concern:

  (i)   The deaths may have been caused by man through neglect,
        carelessness or ignorance.

  (ii)  The deaths may have occurred because an unusual set of
        natural events combined with a previously safe
        agricultural or industrial practice to produce an
        unexpected environmental effect.

  (iii) The deaths may be an indication of a change in the
        environment that could harm man - either physically,
        socially, or economically.

For these 3 reasons, considerable efforts are made to determine
the causes of mortality incidents. This paper primarily describes
work done by the Institute of Terrestrial Ecology when its Monks
Wood Experimental Station investigated the bird mortality
incidents that occurred in the late summer - early winter periods
of 1979, 1980, 1981 and 1982. No notable mortality occurred in
1983.

# WILDLIFE MORTALITY INCIDENTS IN THE UK : GENERAL BACKGROUND

Every year in the UK most types of wildlife are probably affected
by man's activities. Most of the well documented examples of
wildlife mortalities concern deaths of bees, fish, and birds.

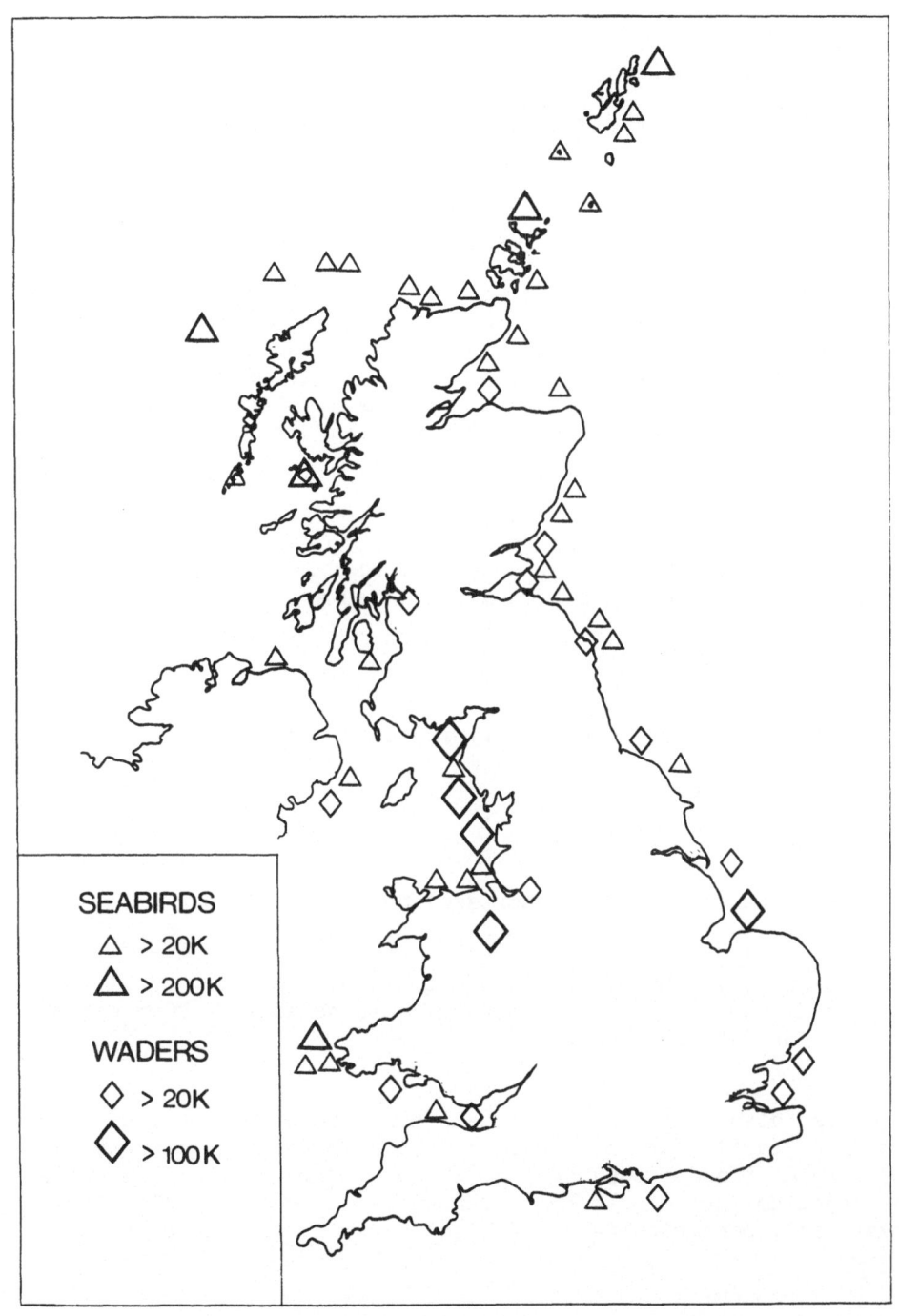

Fig. 1.  U.K. Seabird and Wader Colonies.  (Data based on 1970s figures).

Bee mortality incidents are dealt with by the Agricultural Departments. Sometimes bee mortality incidents occur when bees are accidentally exposed to insecticides sprayed on crops (such as oilseed rape) which are favourite food plants for bees. The economic loss sustained by individual bee-keepers can be substantial.

Fish mortality incidents are investigated and documented primarily by the relevant Water Authority or River Purification Board in collaboration with the agricultural departments. A large mortality can indicate that the waterway is seriously polluted. Water from such a river might be unfit for human consumption, even after purification, so establishing the cause of the incident is of great importance. Industrial or agricultural pollution is often, but not always, the cause of the incident.

A number of organisations are involved in investigating bird mortality incidents in Britain because previous experience has shown that there are many causes of bird mortalities. The five most common causes of mass bird mortalities are:-

1.  Oil.

2.  Pesticides (almost exclusively insecticides + acaricides).

3.  Industrial chemicals.

4.  Disease.

5.  Natural causes (eg food shortage, bad weather).

The possibility that overfishing could be depriving seabirds of food means that a sixth category could be added - "Competition with man for natural resources".

There is also concern in Britain over the deaths of individual birds of particular species, namely those of high conservation value, and those occupying the highest levels of the food chain. Many species are in both categories, eg golden eagle (Aquila chrysaetos), peregrine falcon (Falco peregrinus), and heron (Ardea cinerea).

Notable causes of death of individual birds include:

1. Pesticides (accidental exposure and deliberate or unintentional poisoning).

2. Accidents (collision with buildings and vehicles).

3. Natural causes (eg starvation).


BIRD MORTALITY INCIDENTS : GENERAL INVESTIGATIVE PROCEDURES
IN THE UK

There have been cases of bird mortality incidents of many kinds in Britain since the Second World War (Table 1). During this period bird mortality incidents were first investigated extensively at the Nature Conservancy's (NC) Monks Wood Experimental Station following early work by Ministry of Agriculture scientists, voluntary bodies and independent analysts. Later, the Ministry of Agriculture, Fisheries and Foods (MAFF) set up its own investigative department at Tolworth, Surrey which now routinely monitors wildlife mortality incidents arising from the use (or misuse) of agricultural chemicals. Since becoming part of the Institute of Terrestrial Ecology (ITE) in 1973, Monks Wood Experimental Station has continued to investigate bird mortality incidents. This includes those where industrial chemicals are suspected of causing the incident, as well as the larger incidents involving agricultural chemicals.


Co-operation between MAFF and Monks Wood Experimental Station (whether part of NC or ITE) has always been important when investigations of the larger incidents have been done, and an informal system for investigating the larger incidents has now grown up, which often also involves the Nature Conservancy Council (NCC), the direct descendent of the NC, the Royal Society for the Protection of Birds (RSPB) and the British Trust for Ornithology (BTO). The members of these latter two organisations are often the first to spot and report an incident.


Under this system a typical scenario for the investigation of a bird mortality incident might be as follows. The whole investigation falls into four phases: A. Notification; B. Analysis; C. Research; D. Action on problem.

TABLE 1

Some Example British Bird "Incidents"

| Time | Animals | Chemical | Effect | Investigated by |
|------|---------|----------|--------|-----------------|
| 1950s | Birds of prey | DDE | Population decline | MAFF<br>Monks Wood (NC) |
| 1960s | Birds of prey | Dieldrin | Adult mortality | MAFF<br>Monks Wood (NC) |
| 1969 | Auks<br>(Irish Sea) | PCB?<br>Weather? | Mass mortality (15,000) | Monks Wood (NC)<br>MAFF |
| 1972 | Gannets<br>(Irish Sea) | PCB/DDE/<br>dieldrin | Many dead | Monks Wood (NC) |
| Mid-1970s | Geese<br>(Yorks -<br>E. Anglia) | Carbophenothion | Mass Mortality (2,000+) | MAFF |
| 1979-1982 | Waders<br>(Mersey) | Alkyl lead | Mass mortality (3,000+) | Monks Wood (ITE)<br>NWWA |
| 1975 → | Swans | Lead pellet (in gut!) | Largest cause of death | Monks Wood (ITE)<br>MAFF<br>Oxford University |
| 1982 | Swans | Alcohol | Few drunk | ? |
| 1983 | Auks<br>(North Sea) | None identified | Mass mortality (34,000) | Monks Wood (ITE)<br>MAFF |
| 1983 | Passerines<br>(E. Anglia | Aldicarb | Many dead (500?) | MAFF |

## A. Notification

1. Sick, dead and dying birds noticed by a member of the public (often a member of the RSPB).

2. Member of public informs the local MAFF Veterinary Investigation Centre (VIC) (often indirectly via RSPB HQ at Sandy, or after contacting a local authority, eg a district council environmental health officer).

3. The local VIC carry out preliminary post mortems, especially checking for disease. They dispatch samples to MAFF's central investigations laboratories at Tolworth (agricultural chemicals) and Weybridge (more detailed veterinary checks).

4. RSPB HQ, or MAFF (VIC's or Tolworth) inform NCC and ITE, in case they have not heard independently through their regional representatives of the incident. NCC requests ITE to investigate if it is at all possible that the incident is not just an agricultural chemical that has caused a minor poisoning incident.

   Scientists from MAFF, NCC or ITE often visit the site to confirm the reports and collect special samples when necessary.


## B. Analysis

5. MAFF and ITE laboratories establish contact between the investigating scientists and chemical analysts. The MAFF analysts concentrate on checking for the presence of a wide range of the current agricultural chemicals, while ITE analysts concentrate on organochlorine insecticides (which still occur in British birds although their use is severely restricted), toxic metals and certain industrial products.

6. On the basis of the preliminary analytical results the investigating scientists decide which laboratory should proceed with any subsequent investigation.

7. In many cases, in the early stages of an investigation, relatively few samples need be analysed. If a chemical is the prime cause of an incident it should normally be present in unusual amounts in all the affected individuals - unless of course it breaks down after the animal has died.

## C. Research

8.  Because each incident must be treated as an individual case it is difficult to generalise about how a research programme could be conducted.

9.  A number of incidents are difficult to explain fully and further work is often required both in the field and in the laboratory.  The laboratory work has probably been under-emphasised in many past cases but there have been many cases where laboratory experiments were or could have been (or, indeed, should have been) used to investigate hypotheses established on the basis of the field observations. Laboratory experiments have definitely improved our understanding of the effects of many classes of pollutants. In particular, laboratory experiments were very useful in the investigation of incidents 3, 5 and 6 in Table 1. Laboratory experiments were less than conclusive when investigating into the Irish Sea auk wreck of 1969 were in progress but, at least, they raised a number of questions. Some of these remain unresolved even now.  It must never be forgotten that no matter how good the data is, field observations are in strict scientific terms only "correlative" or "circumstantial".

10. The best experiments would, of course, be those where a quantity of the suspect chemical were actually released into the environment in a controlled fashion and its effects closely monitored.  However, such experiments - effectively attempting to repeat the "ecoaccident" - would raise ethical and legal difficulties in many countries, including the UK.

11. Although it may be difficult or impossible to undertake experiments in the wild (very few have actually been reported) good quality observations can be made.  The value of these observations will be considerably enhanced if they form part of a long running sequence of observations on the ecosystem in question.  Fortunately, in many instances in Britain, a considerable body of background data exists on the levels of pollutants to be normally found in both dead and healthy examples of the genus, if not the actual species, of bird involved in the incident.  There is also often a good deal of national and local knowledge about the ecology of birds.  Thus, there will frequently be data on the normal occurrence, numbers, behaviour, food, reproduction and breeding success of the species involved in the mortality incident.  Not all countries are so fortunate in this

respect. It is also to be remembered that there may well be more information available about the ecology of the animals than there is about toxicity of a suspect chemical to wildlife, or its behaviour in the particular ecosystem concerned.

12. Other organisations than MAFF and ITE often make substantial, if not vital, contributions to the research and analytical stages of wildlife mortality investigations. In some cases, industrial companies contribute analytical expertise, while in other cases, large regional authorities with statutory responsibilities (eg Water Authorities) undertake their own investigations into parts of the problem not being covered by either ITE or MAFF. University and other higher education centres make their expertise and facilities available, but their contribution to this area of studying "ecoaccidents" is perhaps small at present.

13. Ideally the research into any incident will allow, at least, some firm evidence to be gathered as to the cause of the incident. It should also give some idea as to the influence of other environmental pressures on the birds (eg disease).


D.  Action

14. Once the most likely cause of the incident has been identified any action required should be taken by the relevant authority. MAFF itself has some statutory powers in this respect and can initiate action, on the use or method of application of a pesticide. ITE has no such statutory powers and therefore advises the appropriate responsible authorities as to what the conclusions of its research are. Appropriate action is then taken by the relevant authority. Relevant authorities may be regional bodies like Water Authorities, or County Councils, or they may be Government Departments.

15. Needless to say, many months may pass between the first notification and the completion of scientific investigations For this reason the appropriate authorities may have to take action in the absence of a great deal of relevant information and almost always in the absence of complete scientific evidence. The alternative might be to risk a continuation or repeat of the "ecoaccident". While such (scientifically ill-informed) decisions may be unpopular in affected quarters, they will be justified if the intensity of the effects of the ecoaccident diminish when the decision is implemented.

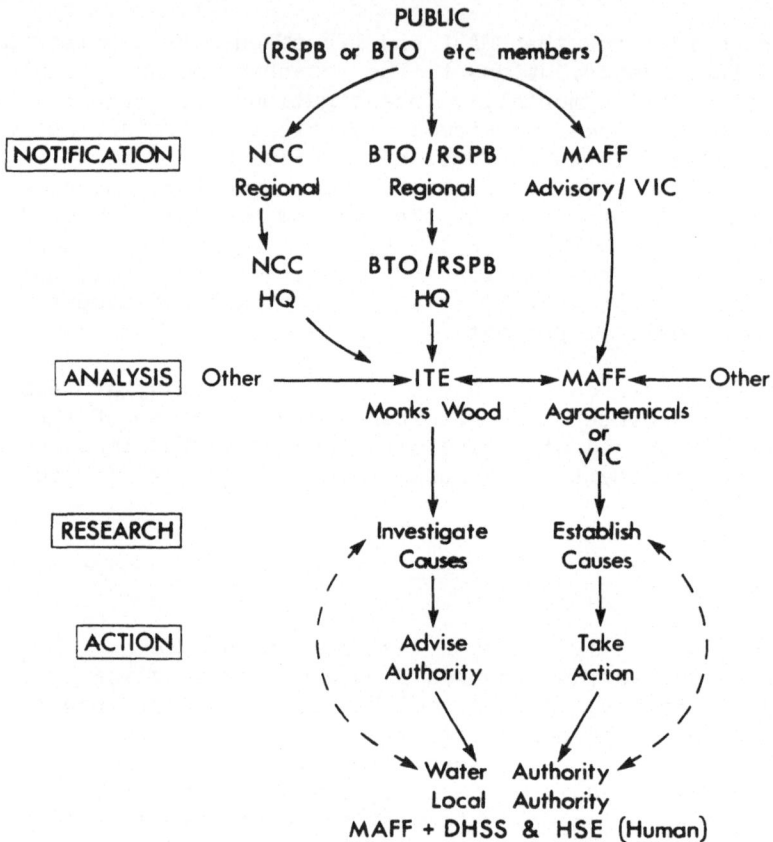

Fig. 2.   The schema for bird incident investigations shown above
          only gives a general outline of the procedure that has
          evolved.   Specific schemes operate for particular species
          or animal groups.   "Other" includes industrial analytical
          laboratories.

A summary of the various steps in the UK incident investigation scheme, as it has evolved to date, are given in Figure 2.


THE MERSEY ESTUARY BIRD MORTALITY INCIDENT


The incident has been fully described in a number of publications, as has the experimental work associated with it (Head, D'Arcy & Osbaldeston 1980; Osborn & Bull 1982; Bull et al. 1983;Osborn et al. 1983; Newton et al. 1983). The main points of the investigation are given below. Readers requiring fuller details should refer to the original publications. As in the typical case given above, the incident is described under the headings: Notification; Analysis; Research and Action; along with an additional heading: Follow-up. In many investigations it has seemed sufficient to take action to solve the problem. The situation is not always examined subsequently to determine the effectiveness of the action pursued in detail.


A. Notification

Following a high tide in the highly industrialised Mersey Estuary in mid-September 1979, two local ornithologists, observing birds on the north shore of the estuary, saw a number of dunlin (Calidris alpina) behaving abnormally. Closer examination showed that there were also a number of dead birds on the tide-line. Over the next few days more birds came ashore and it was clear that gulls and other wading bird species were also involved in the incident. Collection of corpses was initially co-ordinated by staff at Liverpool Polytechnic. Some corpses were passed quite rapidly to the Liverpool MAFF VIC who in turn sent tissues to the appropriate MAFF laboratories for chemical analysis. NCC and ITE did not become involved in the study until the magnitude of this mortality incident became apparent, the next tide bringing the total of corpses seen into the hundreds. Alerted by the RSPB who had been told of the seriousness of the incident by the local ornithologists, the NCC HQ and regional units were soon in contact with ITE scientists at Monks Wood Experimental Station and a selection of carcasses were dispatched to Monks Wood for analysis.

The initial reports conveyed the information that casualty birds were coming ashore dead and sick on both the north and south shores of the estuary, that the birds were unable to fly, that they were not feeding normally, and that they had a most unusual head tremor. In addition, the droppings of sick birds were bright green.

## B. & C.  Analysis and Research

The first batch of birds received at Monks Wood Experimental Station came from the north shore of the estuary. Only five birds were selected as being suitable condition for dissection, the rest being so badly covered in sediment that dissection could not have been done without the risk of contaminating the internal organs on dissection.

No unusual quantities of organochlorine insecticides or PCBs were found in the livers of the five birds that were dissected.

The only unusual analytical finding was that the birds contained high concentrations of lead (Table 2). These values were compared with what are considered as normal liver lead values in these and similar species which had been analysed at Monks Wood since 1963. The lead levels in the livers of the Mersey birds were at least an order of magnitude above those normally found in waders in the UK and it was therefore thought possible that lead was the cause of death.

Figures are mg/kg$^{-1}$ wet wt; ND, none detected. The last bird was badly emaciated and the essential metal levels in the liver were no greater than would be expected in such a bird.

TABLE 2

Toxic chemical concentration in livers of dunlin and redshank found dead on the Mersey

|  | Zn | Cu | Fe | Hg | Cd | Pb | Organochlorine compounds |
|---|---|---|---|---|---|---|---|
| Dunlin | 29.9 | 4.24 | 544 | 2.1 | ND | 18.5 | ND |
|  | 26.3 | 8.00 | 383 | 2.7 | ND | 31.0 | ND |
|  | 22.2 | 6.25 | 370 | 1.4 | ND | 12.8 | ND |
| Redshank | 15.2 | 7.75 | 812 | 0.8 | ND | 7.5 | ND |
|  | 45.2 | 30.0 | 1867 | 4.7 | ND | 20.2 | ND |

A sample of sick and a number of live waders were obtained from the estuaries. These too contained elevated lead levels (Fig. 3).

After the mortality incident in 1979 began a large number of different organisations became involved (Table 3), all of whom had some legitimate interest in the incident, its causes, and its implications. However, because the bird mortality involved, it seemed, an industrial chemical and, also, because it had taken place on an estuary and could involve an industrial effluent, the two public organisations which carried out most of the investigations were ITE and the North West Water Authority (NWWA).

The following paragraphs outline the work done by ITE on the birds; this work was complemented by an extensive monitoring programme of the water and invertebrates of the estuary mud- and sandbanks which was conducted by the NWWA.

TABLE 3

The "experts" with an interest in the Mersey incidents

A   Liverpool Polytechnic
A   Liverpool University Veterinary Centre
*+  MAFF Veterinary Centre (Liverpool)
*   NWWA River Division
*   NWWA Biological Division
*   NWWA Chemical Division
+   MAFF Weybridge Veterinary Centre
φ   Royal Society for Protection of Birds
φ   Cheshire Ornithologists club
+   Nature Conservancy Council
*   Cheshire Local Authority
*   Merseyside Local Authority
φ   Frodsham Wildfowlers' Club
C   Associated Octel
+   Institute of Terrestrial Ecology
φ   British Association for Shooting and Conservation
φ   Rob Cockbain and Malcolm Lord
A   Nuffield Institute (London Zoo)
+   Department of Health
+   MAFF (Food Sciences)
φ   Royal Society for Protection of Cruelty to Animals
+φ  Royal Commission on Environmental Pollution


A = Academic Institute or University
φ = Voluntary Body
+ = Government Institute or Department
* = Local Authority or Regional Authority
C = Company

72

Studies in the field : The mortalities

The 1979 mortality


In 1979, dead and dying birds were first observed on the estuary
in mid-September. From then until early 1980, when the last few
casualties were reported, about 2400 birds were known to have been
affected, the great majority of which died. Details of the
species affected are given in Table 4 (see also Head et al. 1980;
Bull et al. 1983), but most casualties were waders, ducks or
gulls. About half the recorded deaths were for one wader species
namely dunlin (Calidris alpina). Deaths of about 400 black-headed
gulls (Larus ridibundus) were also recorded. It is not known what
proportion the available figures represent of the true number of
affected animals, as many carcases may have been lost to the
river, taken by predators, or washed ashore in inaccessible or
unsearched areas.


Veterinary investigations by other laboratories have been
summarised elsewhere (Head et al. 1980). None of the studies was
able to attribute the deaths to any disease. Similarly, early
post-mortems at Monks Wood eliminated the possibility that food
shortage and subsequent starvation were the cause of death, as
several of the birds found dead had died before their fat and
protein reserves were exhausted. However, these post-mortems did
show that affected birds had discoloured livers, brilliant green
bile, enlarged gall bladders and activated bone marrow. A
consistent finding in dead birds was that lead levels were
unusually high. Sick birds also had elevated levels of lead, and
even healthy birds from the Mersey containd lead concentrations
well above those found in birds on other estuaries (Figure 3).
Furthermore, analysis at Monks Wood of tissues from sick and live
Mersey birds suggested that they had been exposed to a relatively
large acute dose of lead, because kidney levels were close to
those in the liver and other soft tissues (see also Head et al.
1980). In chronic exposure, the kidney would be expected to
accumulate much greater levels of lead than the liver.


Industrial analysts showed that 30-70% of the lead in the tissues
of the birds was present in the form of alkyl lead compounds,
mainly trimethyl lead. Analyses at Monks Wood, using an anodic
stripping voltammetric technique, confirmed this finding.

## TABLE 4

Birds found dead and sick (casualties) on the Mersey estuary, 1979-1982

(The counts were carried out between August and February of the following year, most being completed before mid-December)

| | 1979 | | 1980 | | 1981 | | 1982 | |
|---|---|---|---|---|---|---|---|---|
| | Total dead and sick | % sick | Total dead and sick | % sick | Total dead and sick | % sick | Total dead and sick | % sick |
| **Waders:** | | | | | | | | |
| Dunlin Calidris alpina | 1336 | 19 | 2 | 100 | 5 | 100 | 2 | |
| Redshank Tringa totanus | 116 | 46 | 5 | 60 | 5 | 100 | 2 | |
| Curlew Numenius arquata | 49 | 41 | 8 | | 3 | 100 | | |
| Knot Calidris canutus | 6 | | | | 1 | 100 | | |
| Grey plover Pluvialis squatorola | 5 | | | | | | | |
| Little stint Calidris minuta | 5 | | | | | | | |
| Green sandpiper Tringa ochropus | 4 | | | | | | | |
| Ruff Philomachus pugnax | 2 | | | | | | | |
| Lapwing Vanellus vanellus | 2 | | 4 | 25 | 2 | 100 | | |
| Golden plover Pluvialis apricaria | 1 | | | | | | | |
| Greenshank Tringa nebularia | 1 | | | | | | | |
| Bar-tailed godwit Limosa lapponica | | | 1 | | | | | |
| Common sandpiper Tringa hypoleucos | | | 1 | 100 | | | 1 | |
| Totals or overall % sick | 1527 | 22 | 21 | 33 | 16 | 100 | | |
| **Gulls:** | | | | | | | | |
| Black-headed gull Larus ridibundus | 368 | 2 | 343 | 1 | 11 | 10 | 23 | |
| Herring gull Larus argentatus | 29 | 7 | 87[a] | 2 | 11[a] | 18 | 39 | |
| Common gull Larus canus | 16 | 12 | 22 | | | | 2 | |

| | No. | % sick | No. | % sick | No. | % sick |
|---|---|---|---|---|---|---|
| Lesser black-backed gull Larus fuscus | 4 | | 50 | 4 | 1 | 1 |
| Greater black-backed gull Larus marinus | 1 | | 50 | 4 | 1 | 1 |
| Unidentified gulls | 15 | | 197 | | | 16 |
| **Totals or overall % sick** | 433 | 3 | 699 | 2 | 23 | 14 |
| **Wildfowl:** | | | | | | |
| Teal Anus crecca | 48 | 25 | 1 | 100 | 7 | 71 |
| Pintail Anas acuta | 37 | 40 | 1 | | 2 | |
| Mallard Anas platyrhynchos | 34 | 20 | 30 | | 1 | |
| Shelduck Tadorna tadorna | 16 | 20 | | | 1 | |
| Mute swan Cygnus olor | 2 | | | | | |
| Wigeon Anas penelope | 2 | | | | | |
| Barnacle goose Branta leucopsis | 1 | | | | | |
| **Totals or overall % sick** | 140 | 27 | 31 | 10 | 9 | 67 |
| **Others:** | | | | | | |
| Starling Sturnus vulgaris | 5 | | | | | 14 |
| Heron Ardea cinerea | 2 | | | | | |
| Cormorant Phalacrocorax carbo | 1 | 100 | | | | |
| Moorhen Gallinula chloropus | 1 | | | | | |
| Carrion crow Corvus corone | 1 | | | | | |
| | 10 | 10 | | | | |
| Unidentified birds | 232 | | c113 | | | c25 |
| **Totals or overall % sick** | 2342 | | 864 | | 48 | |

a Type C botulism confirmed in 2 of these gulls in both 1980 and 1981.

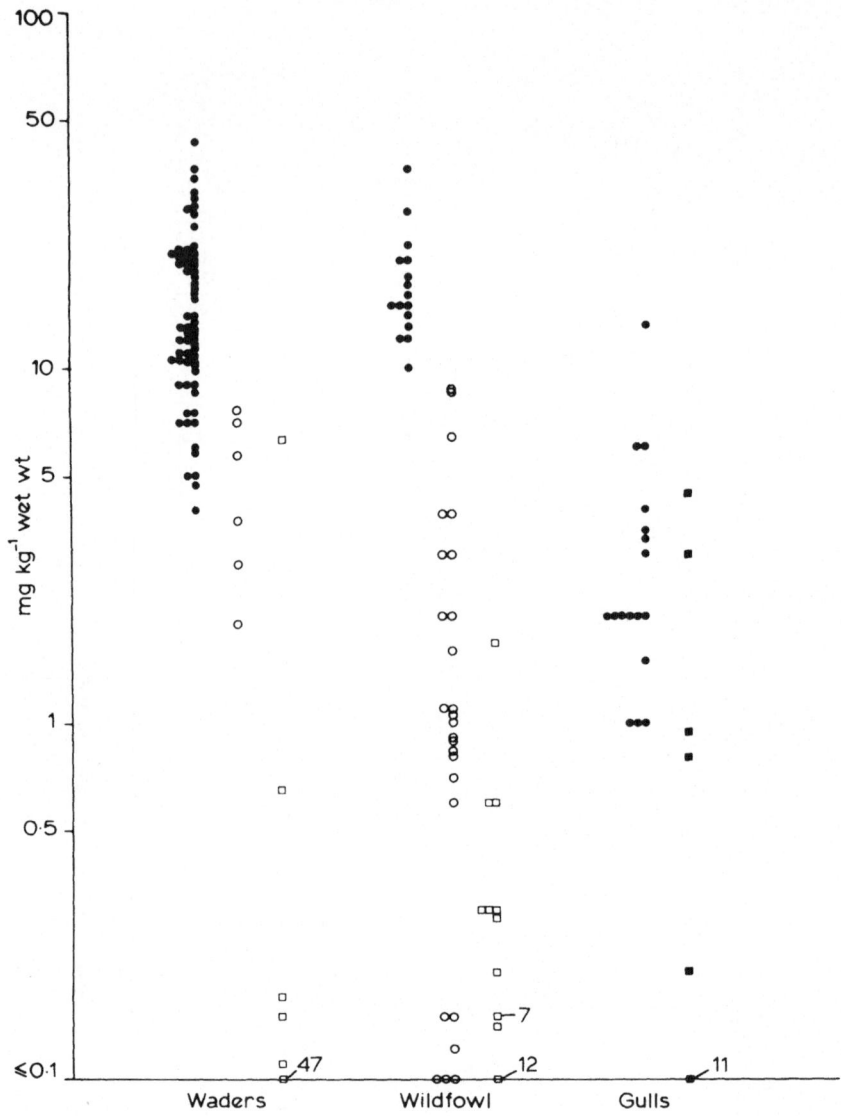

Fig. 3. Liver lead levels (mgkg$^{-1}$ wet wt) in waders, wildfowl and gulls from the Mersey and elsewhere. ●, Mersey casualties; ○, Mersey live caught birds; ■, non-Mersey casualties; □, non-Mersey live caught birds. All birds from the Mersey were collected between September 1979 and December 1980 during the periods when mortalities were occurring or soon after. Birds from other sites were collected specifically for comparison with Mersey data during the same time periods or were collected at earlier times and analysed for lead soon after collection. Not all the data is from ITE analysis. Some of the casualty dunlin and almost all the casualty wildfowl have been analysed by industrial analysts, while MAFF laboratories analysed the Mersey casualty gulls.

The 1980 mortality

At least 850 birds were found dead and dying in 1980, but, in contrast to 1979, black-headed gulls were the major species involved. Unusually low numbers of dunlin seemed to be present at the time of the 1980 incident (see later). Because gulls were such predominant casualties in 1980, the North West Water Authority (NWWA), together with MAFF, investigated the possibility that the birds had died as a result of type C botulism, and 2 herring gulls (<u>Laurus argentatus</u>) were confirmed to have the botulism toxin in their blood.

As in 1979, the waders and ducks affected in 1980 contained high levels of lead, much of which was in the form of alkyl lead compounds. Post-mortem findings once again indicated that affected birds had brilliant green bile, discoloured livers, and green-stained, or discoloured, intestines. It was also noted that affected birds had a distinctive odour.

The 1981 mortality

Very few dead and dying birds were reported in 1981. However, the behaviour of affected ducks and waders was similar to that seen in previous incidents. Post-mortem examination and chemical analysis also produced similar results: affected birds had discoloured livers, green bile, and the distinctive odour. Alkyl lead levels of up to 7 mg Pb/kg (wet weight) were found. Thus, it seemed that alkyl lead compounds were still seriously affecting some birds in autumn 1981.

The 1982 mortality

Once again, the total number of dead birds found on the estuary was very limited. However, one species, the grey heron (<u>Ardea cinerea</u>) was very prominent, perhaps because unusually large numbers of this species (up to 30) were seen feeding near the area where the large 1979 mortality had occurred. About 140 dead birds were reported in from the Mersey area, 23 of which were examined by ITE. Of these birds 7 had alkyl lead levels of 10 mg/kg wet wt and they all had internal features similar to those seen in birds from the 1979 mortality. It was thought these birds had died from alkyl lead poisoning.

Table 4 details the species affected in the incidents.


Studies in the field : Apparently healthy birds


Figure 3 shows that apparently healthy birds shot on the estuary had higher levels of lead than did apparently healthy birds from other estuaries. Since the Mersey is an important wintering area for waders and wildfowl it was necessary to investigate this more fully to see whether these birds were sub-lethally affected by the alkyl lead and were, therefore, "at risk".


Table 5 shows that when apparently healthy birds were examined at post-mortem, several had internal features similar to those found in the casualty birds (enlarged gall bladders, discoloured intestine, brilliant green bile, discoloured liver, and the distinctive odour noted in many of the casualties). This suggested many birds containing more than 1 mg/kg alkyl lead in the liver were at risk from alkyl lead poisoning, assuming of course shot birds do not represent a biased sample of the population.


However, despite this risk and the mortalities in 1979 and subsequent years, the bird counts carried out on the estuary (Table 6) suggest that the total number of birds using the estuary has not changed greatly since the incidents. However, the number of redshank (Tringa totanus) and the number of dunlin in the early part of the season did seem to be lower than in the pre-incident period. If there is a real risk to a large proportion of the birds using the estuary, under the present conditions a great deal of further field research would be needed to quantify the risk.


Botulism


Throughout our investigations the NWWA, in co-operation with the Nuffield Institute for Comparative Zoology in London and the Veterinary service of MAFF, have examined dying birds to check for botulism which seems to affect and kill many birds in the UK in most years. However, confirming that a bird has been affected by botulism - ie demonstrating the botulinus type C toxin in the

# TABLE 5

Tissue concentrations of alkyl lead compounds in some apparently healthy birds collected from the Mersey and elsewhere, and the proportion with internal characteristics similar to dead and sick Mersey birds

| | n | Alkyl lead as mg Pb kg$^{-1}$ (wet tissue) | | | Number with abnormalities | Mean liver alkyl lead (mg Pb kg$^{-1}$) in abnormal birds | Mean liver alkyl lead in normal birds |
|---|---|---|---|---|---|---|---|
| | | Muscle | Liver | Kidney | | | |
| **Wildfowl:** | | | | | | | |
| Mersey | 24 | 1.02±0.19 (0.02-3.42) | 1.90±0.30 (0.02-5.38) | 3.26±0.63 (0.92-13.0) | 18 | 2.34±0.34 (0.2-5.28) | 0.55±0.26 (ND -1.60) |
| non-Mersey | 4 | 0.05±0.02 (0.02-0.1) | 0.03±0.01 (0.02-0.04) | 0.14, n=2 (0.19,0.09) | 0 | | |
| **Waders:** | | | | | | | |
| Mersey | 10 | 0.33±0.08 (0.11-0.97) | 0.72±0.18 (0.17-1.94) | 1.18±0.27 (0.40-2.81) | 4 | 0.79±0.40 (0.23-1.94) | 0.68±0.20 (0.26-1.39) |

Winter numbers of selected duck and wader species on the Mersey estuary: mean for 1971-77 compared with counts for 1979-80, 1980-81 and 1981-82
(G. Thomason and others, pers. comm.)

|  |  | Sep | Oct | Nov | Dec | Jan | Feb | Mar |
|---|---|---|---|---|---|---|---|---|
| Mallard | 1971-77 | 920 | 740 | 910 | 1350 | 1150 | 890 | 500 |
|  | 1979-80 | 310 | 770 | 850 | 1250 | 1750 | 1400 | 560 |
|  | 1980-81 | 1250 | 1700 | 2400 | 1750 | 2400 | 1300 | 660 |
|  | 1981-82 | 190 | 660 | 2300 | 800* | 1200 | 1300 | 800 |
| Teal | 1971-77 | 2400 | 4200 | 6900 | 7100 | 7900 | 6100 | 3400 |
|  | 1979-80 | 460 | 1000 | 1250 | 13000 | 17500 | 14000 | 6800 |
|  | 1980-81 | 2500 | 7200 | 11000 | 18000 | 20000 | 26000 | 13000 |
|  | 1981-82 | 5400 | 7300 | 9400 | 35000 | 8000 | 6100 | 2300 |
| Pintail | 1971-77 | 1200 | 3300 | 6200 | 8500 | 7700 | 5300 | 1250 |
|  | 1979-80 | 110 | 4200 | 4700 | 9800 | 10000 | 2800 | 165 |
|  | 1980-81 | 1950 | 13000 | 18500 | 8000 | 3900 | 12500 | 4000 |
|  | 1981-82 | 260 | 4200 | 11500 | 6000 | 4900 | 2100 | 52 |
| Shelduck | 1971-77 | 180 | 360 | 700 | 1300 | 2300 | 2500 | 2600 |
|  | 1979-80 | 120 | 390 | 2700 | 7400 | 3600 | 4000 | 920 |
|  | 1980-81 | 1000 | 1300 | 8100 | 11000 | 9400 | 12000 | 3900 |
|  | 1981-82 | 360 | 2200 | 12000 | 6500* | 4900 | 4000 | 1850 |
| Dunlin | 1971-77 | 3900 | 11500 | 25000 | 23000 | 26000 | 25000 | 11500 |
|  | 1979-80 | 1800 | 2400 | 22000 | 21000 | 22000 | 29000 | 2500 |
|  | 1980-81 | 63 | 10000 | 21000 | 40000 | 24000 | 31000 | 18500 |
|  | 1981-82 | 720 | 7100 | 18000 | 13500* | 12000 | 1850 | 430 |
| Redshank | 1971-77 | 670 | 860 | 1400 | 960 | 1100 | 870 | 900 |
|  | 1979-80 | 160 | 300 | 260 | 670 | 530 | 480 | 290 |
|  | 1980-81 | 250 | 510 | 670 | 380 | 210 | 1050 | 600 |
|  | 1981-82 | 98 | 780 | 550 | 530 | 460 | 460 | 600 |
| Curlew | 1971-77 | 900 | 520 | 400 | 570 | 480 | 560 | 720 |
|  | 1979-80 | 530 | 210 | 230 | 135 | 220 | 780 | 450 |
|  | 1980-81 | 1250 | 430 | 490 | 91 | 330 | 300 | 880 |
|  | 1981-82 | 810 | 780 | 390 | 145 | 570 | 400 | 120 |

Note:  Because of the potential danger to human health from eating contaminated wildfowl, wildfowling activities were restricted in 1979/80 and 1980/81.  It has been suggested that the lack of shooting may have encouraged more birds than usual to overwinter on the Mersey in 1980/81.

December 1981 count:  Because of the weather conditions, totals for some species were rough estimates, while some areas were not counted, eg Ince Marsh.  Species marked * are believed to have been under-counted.  On an extra count on 27 December 1981, totals included 10700 pintail and 25000 dunlin.

1982 counts:  The low counts for several species, eg dunlin, might be blamed on the bad weather.

blood plasma – is a difficult procedure. To date only a small amount of evidence has emerged to support the view that some of the birds killed on the Mersey between 1979 and 1982 were botulism victims. At the present time ITE remains unconvinced that botulism was the major cause of the mortalities.


## Experimental studies

Even though there may be extensive evidence from field studies to suggest that a particular toxic chemical might be causing a problem, the evidence, strictly speaking, is 'circumstantial' or 'correlative'. The possibility always exists, however remote, that an important environmental factor has not been measured, and that our conclusions would have been different had measurement of the 'missing factor' been included in the research programme. A laboratory-based experimental study overcomes some of the difficulties faced in the real world, enabling us, at least, to test hypotheses in controlled conditions.

Experimental studies were carried out in this investigation to test the hypothesis that alkyl lead had killed the Mersey birds and was placing a great many others at risk. For this hypothesis to be confirmed, alkyl lead compounds would have to be administered to birds, who would then have to exhibit behavioural phenomena similar to the affected Mersey birds, have the same internal lesions and abnormalities as affected birds, and contain similar amounts of lead.

Two experiments were performed with the 2 alkyl lead compounds with which the birds were most likely to come into contact, triethyl lead and trimethyl lead, the latter being the predominant form in the birds. Three different levels of treatment were given to starlings, a convenient and relatively well-known bird for laboratory work: 2 mg trialkyl lead chloride/day, 200 µg trialkyl lead chloride/day, and 0 trialkyl lead chloride/day - the last, of course, being the control. It was hoped that such dosing with trialkyl lead would result in tissue levels close to those found on the Mersey. The higher dose was expected to be lethal in a short period of time, and the low dose was expected to help determine the 'no effect' level of trialkyl lead compounds in birds.

Dosing was ended after 11 doses and birds were examined to see whether their internal morphology had been in any way affected by the treatment. Also, various measures of body condition were taken, eg weight of muscles, liver and kidney, size of fat reserves, etc. During the course of the experiment, the birds had been observed in order to compare their gross behaviour with that of birds on the Mersey, and their daily food consumption had been measured.

Experimental results

1. Tissue levels of trialkyl lead in the high- and low-dose laboratory starlings were similar to those found in substantial numbers of birds on the Mersey in the autumn and late summer periods (Table 7, Figure 3).

2. Morphological changes in the laboratory birds were similar to those found in Mersey birds containing similar levels of trialkyl lead compounds (Table 8), eg green-stained livers, discoloured intestines enlarged gall bladders. In the trimethyl lead experiment, all the low-dose birds had the same characteristic odour found in the Mersey casualties, but fewer birds had this odour in the triethyl lead experiments. In addition, some activation of bone marrow was observed.

   The results suggest that the presence of trialkyl lead compounds cause a number of characteristic internal lesions, all probably deleterious to the animal's welfare. The internal morphological changes to the enterohepatic system may be 'diagnostic' of trialkyl lead poisoning.

3. Behavioural changes in the laboratory starlings are difficult to compare with birds in the wild, but some comparisons can be made. The low-dose birds all retained the capacity to fly in the triethyl experiment, and all low-dose birds in the trimethyl experiment flew apparently normally. The high dose birds in the triethyl experiment became very quiet and fluffed their feathers as if cold. However, some flew briefly until just before death, which was sudden and unheralded by anything but the mildest 'symptoms', although this may not be uncommon in birds.

   The trimethyl high dose birds exhibited a syndrome so disturbing that 4 of the 6 experimental birds had to be killed. This syndrome consisted of head tremors, some shivering, inability to perch, and severe disorientation.

## TABLE 7

Trialkyl lead levels (as milligrams of lead per kilogram wet weight) in tissues of dosed starlings

| Details | Mean | SE | Range |
|---|---|---|---|
| **Triethyl lead experiment** | | | |
| Low-dose starlings | | | |
| Muscle | 1.47 | 0.36 | 0.4-2.9 |
| Liver | 1.07 | 0.2 | 0.6-1.9 |
| Kidney | 1.85 | 0.17 | 1.4-2.5 |
| Bone | 0.19 | 0.02 | 0.1-0.3 |
| Brain | 0.54 | 0.04 | 0.4-0.7 |
| High-dose starlings | | | |
| Muscle | 20.0 | 2.78 | 9.5-30.0 |
| Liver | 40.2 | 9.35 | 14.6-92.3 |
| Kidney | 19.9 | 3.22 | 9.9-2.7 |
| Bone | 6.0 | 1.59 | 1.4-9.6 |
| Brain | 7.3 | 1.27 | 3.8-12.3 |
| **Trimethyl lead experiment** | | | |
| Low-dose starlings | | | |
| Muscle | 3.07 | 0.55 | 1.6-5.3 |
| Liver | 3.70 | 0.56 | 2.1-5.6 |
| Kidney | 5.38 | 0.88 | 2.8-8.0 |
| Bone | 0.39 | 0.12 | 0.1-0.9 |
| Brain | 3.50 | 0.79 | 1.3-6.6 |
| High-dose starlings | | | |
| Muscle | 11.0 | 2.20 | 6.1-19.4 |
| Liver | 32.4 | 4.9 | 318.5-49.7 |
| Kidney | 30.2 | 6.21 | 17.0-57.3 |
| Bone | 4.3 | 1.14 | 0.2-8.5 |
| Brain | 16.7 | 2.65 | 10.0-26.7 |

Levels in control birds were less than 0.1 in all cases.

Low dose = 200µg day$^{-1}$ PbR$_3$Cl
High dose = 2 mg day$^{-1}$ PbR$_3$Cl

## TABLE 8

Scores of morphological changes in trialkyl lead dosed starlings. Figures are sum of scores for all six birds in the group. The range of individual scores is in parentheses

| | Gall bladder | | Bone | | Gut | | Muscle | |
| --- | --- | --- | --- | --- | --- | --- | --- | --- |
| | Triethyl | Trimethyl | Triethyl | Trimethyl | Triethyl | Trimethyl | Triethyl | Trimethyl |
| Controls | 0 | 3.5(1-1.5) | 0 | 9.5(1-2.5) | 0 | 0 | - | 17.25(2.5-3) |
| Low dose | 12(2-2) | 14.5(2-3) | 6(1-1) | 11.5(1-2.5) | 1(0-1) | 7(1-2) | - | 12.5(1.5-2.5) |
| High dose | 21(3-4) | 22(2-4) | 20(2-4) | 17(1-4) | 3(1-2) | 16(1-4) | - | 0 |

Gall bladder score is for enlargement; Bone score is for redness of marrow; Gut score is for discolouration;
Muscle score is for "condition" or "quantity/quality" of muscle.

Except for muscle: 0 = normal; 4 = greatly different from normal.
For muscle: 0 = wasted; 4 = best possible condition.

TABLE 9

Measures of body condition in starlings
(Mean ± SE; n = 6; Range shown for lipid values)

| | Lean dry Body weight (g) | Pectoral muscle (g) | Body lipid (g) | Intramuscular fat (g) |
|---|---|---|---|---|
| **Triethyl experiment** | | | | |
| Controls | 21.5 ± 0.7 | 1.88 ± 0.08 | 2.92 [2-5] | 0.10 [0.04-0.13] |
| Low dose | 21.5 ± 1.0 | 1.85 ± 0.14 | 1.72** [0.7-2.4] | 0.05 [0.02-0.08] |
| High dose | 17.7**± 0.3 | 1.25**± 0.05 | 0.39* [0.3-0.5] | 0.005* [0-0.01] |
| **Trimethyl experiment** | | | | |
| Controls | 20.5 ± 0.8 (21.3 ± 0.3) | 1.98 ± 0.07 (2.03 ± 0.07) | 4.5 (5) [2-7.5] | 0.16 (0.18) [0.05-0.22] |
| Low dose | 19.7 ± 0.4 | 1.75**± 0.09 | 4.5 [2-11.5] | 0.13 [0.06-0.3] |
| High dose | 17.9**± 0.7 | 1.26**± 0.12 | 2.3 [0.5-11] | 0.02* [0-0.13] |

&ast;   Means so marked are significantly different (P<0.05) from controls; Student's t-test.
&ast;&ast; Significantly different if corrected for body size.
    In the trimethyl experiment one of the controls was found to be deformed on dissection. The figures in parentheses are the means without this bird included.

TABLE 10

Organ weights in starlings (means ± SE) dosed with trialkyl lead compounds.

|  | Liver weight (g) | Kidney weight (g) |
|---|---|---|
| **Triethyl experiment** | | |
| Controls | 2.77 ± 0.23 | 0.91 ± 0.05 |
| Low dose | 2.89 ± 0.13 | 0.98 ± 0.03 |
| High dose | 1.78 ± 0.08* | 0.84 ± 0.05 |
| **Trimethyl experiment** | | |
| Controls | 2.80 ± 0.05 | 0.90 ± 0.03 |
| Low dose | 3.23 ± 0.14* | 0.95 ± 0.05 |
| High dose | 1.47 ± 0.13* | 0.61 ± 0.02* |

* Means so marked are significantly different from the controls $p < 0.05$; Student's t-test.
  The values for the deformed bird did not alter the means importantly.

These birds could not have flown. Green droppings were seen in some low-dose birds in both experiments, and in many of the high-dose birds. The trimethyl birds' droppings were more like those of the affected Mersey animals. These behaviour observations add to the view that trialkyl lead compounds caused the Mersey bird mortalities.

4. Feeding records during both of the experiments were interesting in that they revealed that birds dosed with trialkyl lead compounds had disrupted feeding patterns, in comparison with control birds. This effect is of considerable importance for wild birds whose food supply is less certain.

5. Physiological measurements were taken mainly to help determine the 'no effect' level of trialkyl lead, below which birds would be virtually unaffected, and above which they would be subject to serious, if sublethal, effects.

In birds, there are 2 important measures of physiological condition which are of great significance for breeding, survival, and migration, ie the levels of fat and protein reserves. Indices of these levels are provided by determing the bird's total fat content and by the weight of the pectoral muscles. In both experiments, these measures were much reduced in the high dose birds, and some reduction was also observed in the low dose birds. The most marked effect in the low dose groups was the reduction in muscle weight that occurred in the trimethyl experiment, which was probably first evident after 6 doses (Tables 8 and 9).

An additional useful measure of physiological function is organ weight, particularly of the liver and kidney. It was found that kidney weight was reduced in the high dose trimethyl group and liver weight was increased in the low dose trimethyl group. (The liver often enlarges in the presence of a toxic chemical.) No clear effects on liver or kidney weight were seen in the triethyl experiment (Table 10). It is not certain whether these weight changes constitute a deleterious effect.

The physiological observations suggest that, even at quite low levels, trialkyl lead compounds can adversely affect the physiology of birds in such a way as to additionally 'stress' the bird, or reduce its chances of survival directly.

Experimental conclusions

Trimethyl and triethyl lead compounds kill laboratory birds, when tissue levels are similar to those found in dead birds on the Mersey. At levels insufficient to cause death, physiology and internal morphology are affected in ways likely to reduce survival, or to make an animal less able to deal successfully with 'stress', such as food shortage, bad weather, or disease. From the details of the results available it was concluded that these sublethal effects may begin when concentrations in liver, muscle, kidney or brain reach 0.5-1 µg Pb/g of tissue (on a wet weight basis).


## Overall conclusions from field and laboratory work done in connection with the mortality incidents

1. Dead, sick and live birds from the Mersey contain elevated levels of lead in their tissues, compared with those on other estuaries, most of which is in the trialkyl form. Birds killed with alkyl lead compounds in dosing experiments contained similar amounts of lead to the levels found in dead Mersey birds. In addition, the behaviour and internal features of dosed birds were similar to those of dead and sick Mersey birds.

2. We conclude that the death of birds on the Mersey was primarily the results of contamination of the environment with alkyl lead compounds (see also Head et al. 1980), assuming, of course, that waders and ducks do not respond to lead poisoning very differently from starlings.

3. The experimental results also suggest that birds containing more than 0.5 mg Pb/kg (wet weight) as alkyl lead have changed internal and physiological features which will reduce their survival prospects. As many Mersey birds contain this amount of alkyl lead, and as such birds often have some internal features similar to those in both the experimental birds and the sick Mersey birds, it seems reasonable to conclude that many of the thousands of birds using the Mersey estuary - one of Britain's most important overwintering grounds for ducks and waders - were at some risk from the sublethal levels of alkyl lead compounds they contain.

4. Monitoring of alkyl lead compounds in the Mersey areas should continue until acceptable concentrations have been reached in water and biota.

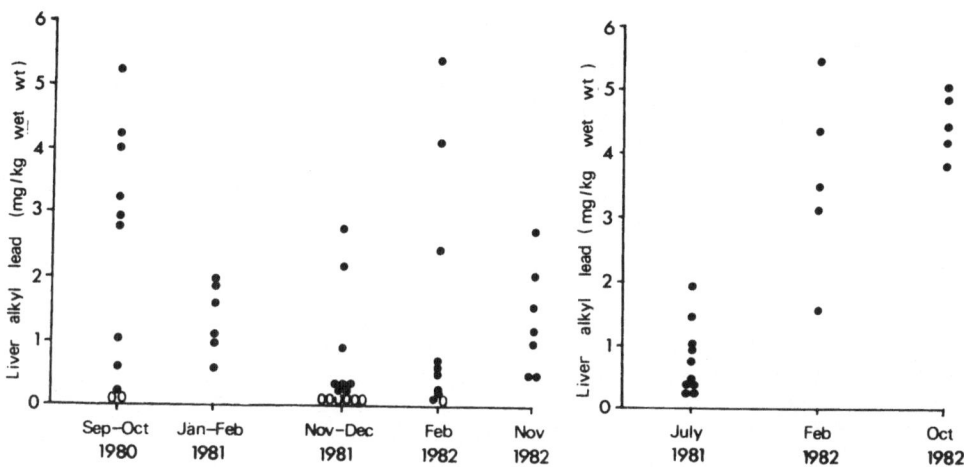

Fig. 4. Concentrations of alkyl lead in livers of live-caught teal (left) and dunlin (right)

Notes:

1. More birds must be analysed before statistical tests can be done or conclusions about trends in the data can be drawn.

2. 0 = <0.1 mg/kg

3. Many of these birds showed internal signs of being affected by alkyl lead (Osborn et al. 1983). In 17 teal examined in February and November 1982, alkyl lead levels in livers of the 10 affected animals were 1.7 mg/kg compared to 1.0 mg/kg for unaffected birds. The respective geometric means were 1.1 and 0.4 mg/kg.

4. Some data were added after the workshop.

5. More recent data is now being analysed. Preliminary results suggest levels in birds are now declining some time after levels of alkyl lead in waterways of the Mersey estuary declined.

## D.  Action taken by relevant authorities

The  results  of the ITE investigations were passed to the NCC and
the  NWWA.    These two bodies have statutory responsibilities for
wildlife  conservation  and  water  quality  control and following
discussions  NWWA  approached industry with a view to reducing the
amount of alkyl lead entering the estuary.

ITE's  results  were also used by health authorities to assess any
threat  to  wildfowlers and their families who were shooting birds
for  food.    Discussions  between  the health authorities and the
British  Association for Shooting and Conservation (BASC) led to a
temporary ban on shooting wildfowl for food.

## Follow-up

As  a  result  of  the  field  and  laboratory  investigations  an
environmental monitoring programme was set up.  This was organised
jointly  by  the  NWWA and ITE.  NWWA measured alkyl lead in water
and  esturine  invertebrates (mainly Macoma) and ITE analysed bird
tissues.    The  purpose  of  the  monitoring  was  to see whether
reductions  in  alkyl  lead  effluent entering the Mersey caused a
reduction  in  the  level  found  in birds.  There are a number of
difficulties  with  "biological  monitoring"  (Moriarty 1984), and
these have been borne in mind.

Figure  4  shows that alkyl lead concentrations in teal - the most
easily  obtained  species  -  have  not  declined greatly over the
period  studied.    Figure    shows that there is little sign of a
decline  in  the  alkyl  lead  in  dunlin either.  The monitoring
programme is likely to continue until some decline is recorded, or
until  we  understand why no decline has occurred.  With good luck
we  shall  learn  a  great  deal about "biological monitoring" from
this study.

An  additional follow up study on the teal was done to see whether
the  decline  in body condition recorded in experimental birds was
also  seen  in  the  field.  Seventeen birds have been examined to
date  but  no evidence to suggest that a decline in body condition

in teal related to tissue levels of alkyl lead compounds has yet emerged (Newton et al. 1983).

Counts of birds on the estuary are also continuing and the low counts of dunlin in the early part of the overwintering season continue to suggest that one particular group of dunlin may have been particularly badly hit by the 1979 mortality.

## A problem resolved?

The absence of any alkyl lead related mortality on the estuary during the autumn of 1983-1984 perhaps suggests that the problem on the Mersey estuary has been resolved.

However, several questions raised by the field and laboratory studies remain unanswered. In this respect, we must await the results of the investigations by the North West Water Authority into the hydrodynamics of alkyl lead compounds in the water and biota of the estuary. It may then be possible to answer the following questions. What is the exact pathway by which the lead reached the birds? Why were mortalities, on the whole, restricted to late summer and autumn? Why, at least initially, were the mortalities associated with high tides? Why were there no noticeable mortalities prior to 1979, and why were they less severe in 1981?

Certainly, data available in the preliminary report from the Water Authority (Head et al. 1980) suggest that sufficient alkyl lead compounds enter the Mersey-Manchester Ship Canal system to account for the levels found. It is likely that prey items for the birds could easily accumulate enough lead from the water for toxic concentrations to occur in birds eating such contaminated prey (Macoma contained about 1 mg / kg lead, mostly alkyl lead compounds). The pathway, therefore, from the alkyl lead effluent outfalls to the birds seems easily explained.

Equally, reductions in alkyl lead effluent may explain why there were fewer mortalities in 1981 and 1982, and no mortalities in 1983.

However, the other questions raised above are more difficult to answer, and depend in part upon unravelling the complex hydrology of the Mersey-Manchester Ship Canal system. Probably, we shall never fully understand why mortalities were not seen before 1979, even though the companies using and manufacturing alkyl lead had been in operation for many years previously. Ironically, it could be that, as the Mersey has become less polluted, the invertebrate and plant foods of these birds have increased, so attracting them to feed in contaminated areas which were previously avoided. If this is so, then the Mersey incident may be the first instance where a general decrease in pollution has led to increased wildlife mortalities, although, of course, the incident was caused by the continued presence of harmful quantities of a specific pollutant.

## General conclusions

1.  Birds died in large numbers on the Mersey estuary because they were poisoned by alkyl lead compounds.

2.  The combination of field and laboratory investigations allowed us to establish the cause of death of the birds, to assess the threat to birds using the estuary, and to recommend an "environmentally safe limit" on the concentration of alkyl lead levels that needed to be achieved in birds to ensure only minimal bird casualties occurred. Without both laboratory and field investigations none of these things could have been done.

3.  Reductions in effluent seems to have progressively reduced the incidence of mortality, even though biological monitoring of birds has not suggested that levels are declining in teal.

4.  Despite the apparent resolutions of the problem a number of unanswered questions still remain. Many of these will probably be answered when studies by the NWWA are complete.

## Acknowledgements

ITE is grateful to many members of other organisations who took part in the study contributing time, advice and expertise. These organisations include the North West Water Authority, Associated Octel Ltd, Ministry of Agriculture, Fisheries and Food, the British Association for Shooting and Conservation, and the Royal Society for the Protection of Birds. This work was funded by the

Nature Conservancy council as part of its programme of research
into nature conservation.

REFERENCES

BULL, K.R., EVERY, W.J., FREESTONE, P., HALL, J.R. OSBORN, D., COOKE,
A.S. & STOWE, T. 1983.  Alkyl lead pollution and bird
mortalities on the Mersey estuary, UK, 1979-1981.  Environ.
Pollut. A., 31, 239-259.

HEAD, P.C., d'ARCY, B.J. & OSBALDESTON, P.J. 1980.  The Mersey
estuary bird mortality autumn-winter 1979 - Preliminary
report.  North West Water.  Directorate of Scientific Services.
Scientific Report Ref. No. DSS-EST-80-1.

MORIARTY, F. 1984.  Ecotoxicology: The study of pollutants in ecosystems.  London: Academic.

NEWTON, I., HASS, M.B., OSBORN, D., BELL, A.A., EVERY, W.J. HANSON,
H.M., HARRIS, M.P. & WYLLIE, I. 1983.  Birds of Prey and Pollu-
tion.  ITE Project 181.  (ITE Interim report to the Nature
Conservancy Council).

OSBORN, D. & BULL, K.R. 1982.  Mersey bird mortalities 1979-1981:
a pollution problem resolved?  Annu. Rep. Inst. terr. Ecol.
1981, 28-33.

OSBORN, D., EVERY, W.J. & BULL, K.R. 1983.  The toxicity of trialkyl
lead compounds to birds.  Environ. Pollut. A, 31, 261-275.

SYNOPSIS OF CASE HISTORIES

D. Calamari

Institute of Agricultural Entomology
University of Milan
Italy

Participants of the workshop decided to prepare a synopsis
that would identify common aspects of the different ecoaccident
cases described (an ecoaccident could be defined as an event or
a series of events unexpected in outbreak and/or effects) and would
expand the discussion to a list of key points that could help in
programming prevention and intervention in future accidents.  A
matrix was considered the most appropriate method for accomplishing
this task.

After each workshop presentation, a number of key points was
selected that provide evidence for the most relevant characters
of each case.  These points were debated by the participants and
relevant items were used for the preparation of the matrix after
a consensus of approval was reached.  In addition, a list of the
relevant scientific problems raised by each case was compiled.

Table 1 contains a matrix showing a high number and variability of possible parameters even for a small number of cases.
For example, in point 3, the ecosystems involved in the accident
are characterized by four different biotopes.  In point 6, the
time of release varies from minutes to years and so on.  Consequently,
this evidence indicates that there are no typical ecoaccidents.
However, compiling this matrix allowed exchange of information and
experiences from the variety of ecoaccidents and demonstrated that
a list of parameters could be identified and considered as the base
for work on the "taxonomy" of ecoaccidents.  If further studies
were based on a higher number of cases, a classification scheme
could be developed and prevention and intervention would be greatly
improved.

## Table 1. Matrix of Case Histories

| | Parameter | Harbour Tange | Otranto Channel | Seveso Area | Mersey Estuary |
|---|---|---|---|---|---|
| 1. | Cause, at the moment of the accident | known | known | known | unknown |
| 2. | Type of accident | leaching from solid waste | transport accident | unwanted industrial gas release | industrial waste effluent |
| 3. | Ecosystem involved | coastal waters | marine waters | land area | estuary |
| 4. | Chemical substances | organo P, organic Hg, several chemicals | tetraethyl lead (TEL), tetramethyl lead (TML) | tetrachlorodibenzo-dioxine (TCDD) | alkyl lead (late identification) |
| 5. | Scientific information available | basic background | not existing | scarce | basic background |
| 6. | Time of release | long (years) | predicted long | short (minutes) | relatively long (days) |
| 7. | Type of release | slow leaching | slow spill or breakage | cloud (quick) | continuous point source |
| 8. | Quantity | evaluation a posteriori | exactly known | estimation a posteriori | estimation a posteriori |
| 9. | Analytical problems | moderately difficult | difficult (type of chemicals) | difficult (low levels) | moderately difficult |
| 10. | Type of effect | fish and bird kills, Hg in fish, loss of diversity | possible effects on marine organisms, moderate bioaccumulation | domestic animal mortality, human health in danger, wildlife bioaccumulation | bird kills, bird populations in danger |
| 11. | Diffusion, leaching | expected | blocked | effective | effective |
| 12. | Residual contamination | remaining a waste at slow dispersion | no residual | remaining contamination at slow dispersion | dispersion |
| 13. | Modelling for prediction | very useful for future prediction | very useful, provoked removal | useful for management | not important |
| 14. | Type of model | environmental distribution and fate, applicable | mathematical model of dispersion, applicable | environmental distribution and fate, scarce data | environmental distribution and fate, explanatory |
| 15. | Safety levels | maximum allowed level of accumulation in biota | not applicable | to be indicated, very difficult | maximum allowed level in the effluent |
| 16. | Action taken | late action taken, partial removal | removed | decontamination in progress | effluent control |
| 17. | Monitoring | operating | ended | operating | operating |
| 18. | Recovery | predicted 30 years | cleaned | very long, not predictable | under control, in fast progress |

Preparation of a registry for this type of event should be assigned to an appropriate, preferably international, organization. The practicality of this approach is, in a certain sense, already demonstrated by experience gained in oil spills. Oil spills have not been taken into account in the present matrix as they already have their own classification due to their specificity and to the high number of events that have occurred (see J. Baker, this volume). In fact, the number of available cases could be used to prepare a flow scheme for examining accidents to develop, for example, a "vulnerability index" for a particular ecosystem in relation to an oil spill. The concept of "vulnerability index" could also be applied to other types of ecoaccidents and be of assistance in preparing maps of risk for given areas. Parameters considered for calculation of a "vulnerability index" might be those in the matrix including a few more (i.e., management of the intervention, clean-up consequences) together with deterministic and stochastic considerations. Moreover, critical examination of case history data and an attentive overlook on the action taken revealed the most important aspect of each accident and led to the identification of important scientific findings, weaknesses in the scientific approach, restriction in the circulation of information, and evidence on problems still unsolved.

At least three unexpected facts were discovered in the Harboør Tange accident: a long persistence of parathion, an apparent lack of speciation of Hg, and an unexpected and highly variable mobility of a number of toxic chemicals via the underground into the coastal ecosystem that led to an unpredictable series of intoxications of the benthic fauna. Evaluation of residual quantities of individual chemical substances in solid waste depots is still an open question. This has led to a certain amount of difficulties in preparing the prediction model for calculation of the speed of leaching and time of recovery.

The most relevant difficulties found during the investigation on the Ontranto Channel were of analytical order. The low level of lead in marine waters and the particular type of chemical substances involved resulted in very slow progress, especially in the initial phase of the investigation. Moreover, background data on the lead level in marine waters were not reliable, and the information on toxicity and bioaccumulation on aquatic organisms of tetraethyl-lead (TEL) and tetramethyl-lead (TML) was completely lacking. The possibility of organolead being converted to methylate mercury was another point to take into account for the assessment of the hazard but the kinetic processes were not known.

The peculiarity of the Seveso accident, in respect to the other case histories, was the possibility of danger to human health. The sudden mass mortality of domestic animals necessitated strong

immediate action.  This haste resulted in an inadequate sampling
scheme at the outset, and only after a certain time did a better
sampling program emerge.  The final fate of tetrachlorodibenzo-
dioxine (TCDD), by natural processes or by disposal or destruction,
caused a great debate, and the problem has not yet been completely
solved.  Another important and still open question concerns the
recovery phase and the possibility of repopulating the evacuated
area.  The toxicological problem is whether or not it is possible
to define a maximum allowable concentration in vegetables and soil,
and, in the case of a positive answer, which level could be considered safe for the human population and the ecosystem.  It has also
been noted that evaluation of the quantity of TCDD released during
the Seveso accident was not carried out with the necessary speed,
although this should have been a major priority.  A reliable common
estimate was obtained only after several attempts.

The major constraint in the Mersey Estuary accident was apparently the slow circulation of available scientific information
among different operative groups.

Consideration must be given to what has been learned from
the accidents described.  In general, with only a few exceptions,
field data confirmed laboratory experiments if they were performed
in comparable conditions.  The importance of basic information has
been stressed by all the lecturers and confirms the necessity of
ecological studies and ecotoxicological testing.  Ecotoxicological
tests for new chemicals are now required by the Toxic Substances
Control Act in the United States and the Directive on Dangerous
Substances in the Economic European Community.  In addition, the
relevance of appropriate sampling schemes and methods for mass
balance has also been stressed, and the usefulness of physico-
chemical models for prediction of distribution and fate of chemical
substances has been widely demonstrated.

A key point for assessing damage caused by an ecoaccident is
recovery time, but very little progress has been made in this area
as all examples that were presented were forecasts or model predictions that still must be verified.  Other aspects that were
considered during the discussion concerned methods of intervention.

Maximum rapidity in giving the alarm of an ecoaccident is to
be recommended.  However, the alarm may be delayed for several
social considerations, such as negative publicity, public opinion,
panic reactions etc.  Immediate action in all the accidents would
have improved scientific investigations and the effectiveness of
the actions taken.  Two tasks should always be undertaken immediately: mapping of the risk area; reliable evaluation of the
quantity of chemical substances released.

A close coordination and a clear sharing of responsibilities among scientists and administrators in their various functions as authorities and advisers is essential. Maximum circulation of information and scientific data, including a well-considered flow of public information, is also recommended because social constraints and local conflicts between interested parties may increase the natural difficulties of the scientific problems in most cases.

# CAUSES AND CLASSIFICATION OF OIL SPILLS

Jenifer M. Baker

Field Studies Council
Preston Montford
Shrewsbury, U.K.

## INTRODUCTION

Compared with other pollutants, the volume of oil (petroleum hydrocarbons) accidentally spilt each year is high, and many case histories have been described both in the scientific literature and the popular press. However, considering the world as a whole, accidental spills, as compared with routine or operational discharges, account for only a minor proportion of the oil entering the environment (as indicated by Table 1).

Table 1. Inputs of petroleum hydrocarbons in the marine environment (million metric tonnes per annum)

| | Best estimate | Probable range | 1973 estimate |
|---|---|---|---|
| Transportation | 1.49 | 1.00-2.60 | 2.1 |
| Tanker operation | 0.71 | 0.44-1.45 | |
| Drydocking | 0.03 | 0.02-0.05 | |
| Marine terminals | 0.02 | 0.01-0.03 | |
| Bilge and fuel oil | 0.32 | 0.16-0.60 | |
| Tanker accidents | 0.39 | 0.35-0.43 | |
| Non-tanker accidents | 0.02 | 0.02-0.04 | |
| Production platforms | 0.05 | 0.04-0.07 | 0.08 |
| Atmospheric | 0.30 | 0.05-0.50 | 0.60 |
| Municipal, industrial wastes, run-off | 1.40 | 0.70-2.80 | 2.80 |
| Natural seeps/erosion | 0.30 | 0.03-2.60 | 0.60 |
| Total | 3.6 | 1.8-8.6 | 6.1 |

Source IMCO (now IMO) 1981

Compared with chronic effluents, accidental spills are
likely to result in high local concentrations of oil, of particular
concern from the ecological point of view.  Sufficient case history
information is available to classify these accidental spills in
various ways.  Such classifications are useful in highlighting
priority areas for response and regulation, and are likely to have
some relevance to ecoaccidents other than oil spills.  Some of the
classifications used in this paper are based on spill characteristics (e.g. cause of spill, type of oil involved) and some on
environmental characteristics which describe in various ways the
nature of the areas receiving the oil.  The examples given here
have been chosen for their diversity within the limitations imposed
by availability of information.  The literature is dominated by
North American and European marine and coastal information.


SOURCES AND CAUSES OF ACCIDENTAL OIL SPILLS

The main sources of accidental oil spills are listed in
Table 2.


Table 2.  Sources of oil spills

Blow-outs
Tankers: groundings and collisions
Tankers: operational accidents
Accidents to ships other than tankers
Pipeline failures
Storage tank and bund failures
Road and rail accidents
Effluent treatment system failures


More detailed analyses have been published for particular
spill sources or geographical areas, and three examples follow,
namely world oil spills from tankers, European pipeline failures,
and oil spills in Nigeria.

## Analysis of world oil spills from tankers

Table 3 is based upon data from the International Tanker
Owners Pollution Federation Ltd. oil spill data collection scheme.
It shows that most accidental spills from tankers occur during
routine operations and are small (less than 50 barrels).  Few spills
result from collisions or groundings, but a high percentage of
these are large (over 5000 barrels).

Table 3. World oil spills from tankers, by cause/operation in progress, 1974-1982

Quantity of oil spilt (barrels)

| | Trace | 0-0.5 | 0.5-5 | 5-50 | 50-5000 | 5000+ | Total | % |
|---|---|---|---|---|---|---|---|---|
| Collision | 10 | 7 | 5 | 13 | 103 | 46 | 184 | 3.9 |
| Grounding | 9 | 6 | 19 | 27 | 112 | 65 | 238 | 4.9 |
| Ballasting/ deballasting | 125 | 120 | 252 | 76 | 23 | 0 | 596 | 12.1 |
| Loading/ discharging | 456 | 451 | 792 | 427 | 201 | 10 | 2337 | 48.0 |
| Tank cleaning | 14 | 10 | 24 | 8 | 7 | 0 | 63 | 1.3 |
| Bunkering | 55 | 121 | 182 | 71 | 21 | 0 | 450 | 9.2 |
| Bilge pumping | 24 | 22 | 39 | 12 | 3 | 0 | 100 | 2.1 |
| Internal transfer | 24 | 31 | 38 | 24 | 5 | 0 | 122 | 2.5 |
| Unknown | 169 | 125 | 191 | 106 | 123 | 68 | 782 | 16.0 |
| Total | 886 | 893 | 1542 | 764 | 598 | 189 | 4872 | |

Source International Tanker Owners Pollution Federation Limited

The causes listed in Table 3 can often be traced to more basic problems. An example of a possible chain of events is:

Spill during discharging
↑
Hose failure
↑
Inadequate maintenance
↑
Inadequate training or other
'human error' factors such as
tiredness

Comprehensive reliable data on basic causes are unobtainable, but human error in its many forms is believed to be of primary importance (see, for example, discussion in Wardley-Smith 1983).

Analysis of European pipeline failures

The CONCAWE classification of pipeline failures (Table 4) presents a different picture from tanker accidents, with external corrosion standing out as the main problem. However, human error expressed as 'accidental third party activity' is the second most important cause of failure, and other types of human inadequacy may of course contribute to the external corrosion and materials failure problems. Moreover, external corrosion and materials failure are probably contributing factors in some of the tanker spills classified in Table 3. Thus some of the common features between tanker spills and pipeline spills may be obscured by the nature of the classification schemes used in Tables 3 and 4.

Table 4.  Causes of western European pipeline failures, 1978-1982

Number of incidents

|  | 1978 | 1979 | 1980 | 1981 | 1982 | 1978-1982 |
|---|---|---|---|---|---|---|
| Mechanical failure | | | | | | |
| Construction | – | 2 | – | – | 1 | 3 |
| Material | 3 | – | 2 | 3 | 1 | 9 |
| Operational error | | | | | | |
| System | – | – | – | – | – | – |
| Human | – | – | – | – | – | – |
| Corrosion | | | | | | |
| External | 7 | 4 | 3 | 7 | 3 | 24 |
| Internal | – | – | – | – | 3 | 3 |
| Natural hazard | | | | | | |
| Subsidence | 1 | – | 1 | 1 | – | 3 |
| Flooding | – | – | – | – | – | – |
| Other | – | – | – | 1 | – | 1 |
| Third Party Activity | | | | | | |
| Accidental | 4 | 3 | 3 | 3 | 1 | 14 |
| Malicious | – | 1 | 1 | – | – | 2 |
| Incidental | – | – | – | 1 | 1 | 2 |
| Total | 15 | 10 | 10 | 16 | 10 | 61 |

Source CONCAWE 1983

## Analysis of Nigerian oil spills (all types)

Table 5 classifies all types of Nigerian oil spills (from pipelines, tanks and ships) by cause.  Equipment failure is the major cause, however this category includes a variety of items such as pipeline corrosion and bursts, tank overflows, and valve and hose failures.  Though operator/maintenance error appears as a separate category in Table 5, it is possible that these human factors also contribute to the 'equipment failure' category.  Sand cut (erosion) and sabotage are pipeline problems of regional concern.

Table 5.  Causes of crude oil spills in Nigeria 1976-1980

| Causes | 1976 No. of spills | % | 1977 No. of spills | % | 1978 No. of spills | % | 1979 No. of spills | % | 1980 No. of spills | % |
|---|---|---|---|---|---|---|---|---|---|---|
| Unknown | 10 | 7.8 | 3 | 2.9 | 13 | 8.4 | 20 | 12.7 | 10 | 4.1 |
| Blowout | 3 | 2.3 | 1 | 1.0 | 3 | 1.95 | 5 | 3.2 | 4 | 1.7 |
| Equipment failure | 63 | 49.2 | 61 | 58.7 | 76 | 49.4 | 83 | 53.0 | 114 | 47.3 |
| Operator/ maintenance error (human) | 21 | 16.4 | 20 | 19.2 | 20 | 13 | 16 | 10.2 | 29 | 12.0 |
| Engineering error | - | - | - | - | 1 | 0.65 | 2 | 1.3 | - | - |
| Natural causes | 1 | 0.8 | - | - | - | - | - | - | - | - |
| Third party | - | - | - | - | - | - | 3 | 1.9 | 6 | 2.5 |
| Sabotage | 30 | 23.4 | 19 | 18.3 | 38 | 24.7 | 25 | 16.0 | 67 | 27.8 |
| Sand cut | - | - | - | - | 3 | 1.95 | 3 | 1.9 | 5 | 2.1 |
| Accident | - | - | - | - | - | - | - | - | 6 | 2.5 |
| Total | 128 | 100 | 104 | 100 | 154 | 100 | 157 | 100 | 241 | 100 |

Source Awobajo 1983

SIZES OF OIL SPILLS

Classification of spills by size shows that most are small. This is clear from statistics for world tanker spills (Table 3), Nigerian crude oil spills (Table 6), and oil spills in and around United States waters (USCG 1983, not reproduced here).  Table 6 illustrates particularly clearly the fact that the total volume of oil spilt may be dominated by a few incidents.

Table 6.  Sizes of crude oil spills in Nigeria 1976-1980

| Spill size (barrels) | Number of incidents | % | Quantity spilled (barrels) | % |
|---|---|---|---|---|
| 0 -  50 | 593 | 75.6 | 6,436 | 0.4 |
| 51 -  100 | 47 | 6.0 | 3,751 | 0.2 |
| 101 -  500 | 80 | 10.2 | 22,808 | 1.2 |
| 501 -  2,000 | 37 | 4.7 | 35,126 | 1.9 |
| 2,001 -  9,000 | 14 | 1.8 | 62,665 | 3.4 |
| 9,001 -  10,000 | 2 | 0.3 | 20,000 | 1.1 |
| 10,001 - 600,000 | 11 | 1.4 | 1,692,261 | 91.8 |
| Total | 784 | 100 | 1,843,047 | 100 |

Source Awobajo 1983

## TYPE OF OIL SPILT

Both crude oil and its products may be spilt. Crude oils from different fields vary considerably in physical and chemical properties, and products (in order of increasing specific gravity) include gasoline, kerosine, gas oils, and light, medium and heavy fuel oils.

Considering the world tanker traffic as a whole, crude oils of various types are most likely to be spilt, but regional coastal tanker traffic is likely to carry more diverse cargoes. Statistics for world tanker spills are given in Table 7.

Table 7. World oil spills from tankers: number of spills of particular oil type, 1974-1982

| Oil type | Quantity | | | | | | | | |
| --- | --- | --- | --- | --- | --- | --- | --- | --- | --- |
| | A | B | C | D | E | F | Unknown | Total | % |
| Crude | 263 | 130 | 543 | 192 | 179 | 90 | 451 | 1848 | 25.4 |
| Bunker | 93 | 156 | 225 | 97 | 65 | 12 | 133 | 781 | 10.7 |
| Bilges | 59 | 45 | 75 | 24 | 6 | 0 | 54 | 263 | 3.6 |
| Fuel (cargo) | 99 | 155 | 239 | 153 | 137 | 30 | 192 | 1005 | 13.8 |
| White product | 78 | 80 | 114 | 63 | 63 | 24 | 109 | 531 | 7.3 |
| Tank washing | 58 | 45 | 111 | 20 | 12 | 0 | 109 | 355 | 4.9 |
| Lube oil | 62 | 35 | 33 | 9 | 6 | 2 | 43 | 190 | 2.6 |
| Bitumen | 5 | 12 | 11 | 8 | 6 | 5 | 8 | 55 | 0.8 |
| Other | 165 | 108 | 163 | 89 | 66 | 15 | 1642 | 2248 | 30.9 |
| Total | 882 | 766 | 1514 | 655 | 540 | 178 | 2741 | 7276 | |

Key for quantity spilt (barrels)

A Trace    D 5-50
B 0-0.5    E 50-5000
C 0.5-5    F 5000+

Source International Tanker Owners Pollution Federation Limited

Pipelines in oil producing areas will carry mostly crude oil, whereas in areas of high consumption such as western Europe, pipeline contents are more diverse. The 10 pipeline spills analysed by CONCAWE (1983) could be classified as follows in Table 8.

Table 8.  Types of oil spilt from western European pipelines 1982

| Oil type | Number of incidents |
|---|---|
| Crude | 5 |
| Light crude | 1 |
| Gasoline | 1 |
| Gas oil | 2 |
| Fuel oil | 1 |

Source CONCAWE 1983

CLEAN-UP

The various methods available for oil spill clean-up may be listed as follows (source Baker 1983).  These methods have been largely developed for coastal or marine use, but some e.g. various forms of mechanical or manual collection, are also appropriate for freshwater and terrestrial spills.  Herding, gelling, sinking and enhanced biodegradation are relatively ineffective, (and expensive) especially if large amounts of oil are involved.

1.  Natural clean-up.  The 'do nothing' approach.  A valid option in many cases, e.g. in some open sea areas remote from bird colonies; on exposed rocky shores; or on marshes where any form of clean-up would result in unacceptable physical disturbance such as trampling damage.

2.  Mechanical.

1. Containment and recovery from the water surface, e.g. using booms and skimmers, or sorbents and various collection devices.  Limitations include sea state and current speeds.

2. Removal by pumping.

3. Use of heavy earth-moving equipment, e.g. on sandy beaches.

4. Use of custom-built beach-cleaning machine.

5. Hosing (high or low pressure) of beaches.

6. Hot water/steam treatment of beaches.

7. Use of cutting equipment (e.g. on marsh vegetation).

8. 'Disturbance' of sediments to release oil, e.g. using high pressure hosing or ploughing techniques.

3.  Chemical.

1. Dispersants. These may provide the only means of dealing with slicks in comparatively rough open sea conditions, and may also be used in nearshore waters and on the shore. However there are several limitations (Baker 1983).

2. 'Herding' using surface-active agents.

3. Gelling.

4.  Manual.

1. Collection of tarballs.

2. Handspreading and collection of sorbents such as straw.

3. Cutting, e.g. oily seaweed or reeds.

5.  Sinking. Sinking of slicks of relatively heavy oils, e.g. with sand. This should not be done where there is any likelihood that oil will reach vulnerable bottom areas such as shellfish beds.

6.  Burning. Burning of floating or stranded oil is usually difficult because of rapid loss of volatiles; however this technique seems promising for polar regions (Meikle 1981) and air-droppable igniters for remote areas have been developed.

7.  Enhanced biodegradation. Biodegradation of oil occurs naturally; oil-degrading organisms are ubiquitous and rapidly multiply under suitable conditions. The degradation process requires oxygen and nutrients. Under some circumstances bio-degradation may be enhanced through addition of micro-organisms and/or nutrients and oxygenation. Oxygenation is promoted by dispersal of the oil, which increases its surface area.

Classification by clean-up approach goes hand in hand with environmental classifications to produce oil spill contingency plans. An essential feature of these is that areas under threat of oil pollution are, before the event, classified (usually mapped), and appropriate clean-up methods for each environmental category are decided.

# ENVIRONMENTAL CLASSIFICATIONS

Environmental classifications may include organisms (Table 9), habitats (Table 10), or categories fulfilling local or regional practical requirements (for example, Tables 11, 12, and 13).

Table 9.  Ecological effects of oil spills and clean-up:
          organisms to be considered

> Mammals
> Birds
> Reptiles and amphibians
> Fish
> Invertebrates
> Vascular plants
> Algae

Table 10. Ecological effects of oil spills and clean-up:
          habitats to be considered

> The open sea
> The seabed
> Coral reefs
> Intertidal rocks
> Mud and sand flats
> Saltmarshes
> Mangrove swamps
> Freshwater habitats
> Terrestrial habitats

Table 11. Classification of coastal types for the
          purposes of 'a field guide to coastal oil
          spill control and clean-up techniques'
          CONCAWE 1981

> Sandy beach
> Pebble or shingle beach
> Boulders
> Scattered rocks and tidal pools
> Cliffs
> Salt marshes
> Muddy coasts
> Estuaries
> Small islands and deltas

Table 11.  Continued

Fjords
Amenity beach with installations
Marinas
Fishery harbors
Industrial ports
Industrial coastal installations

Table 12.  Environmental distribution of Nigerian
oil spills 1978-1980

| Environ- mental types | 1978 | | | | 1979 | | | | 1980 | | | |
|---|---|---|---|---|---|---|---|---|---|---|---|---|
| | No. of spills | % | Net vol.* | % | No. of spills | % | Net vol.* | % | No. of spills | % | Net vol.* | % |
| Inland waters | 9 | 5.8 | 790 | 0.8 | 11 | 7.0 | 3984 | 0.6 | 28 | 11.6 | 1356 | 0.2 |
| Swamps | 76 | 49.4 | 24896 | 25.6 | 63 | 40.1 | 560331 | 88.9 | 70 | 29.1 | 29715 | 5.3 |
| Offshore | 21 | 14.3 | 68795 | 70.7 | 26 | 16.6 | 31440 | 5.0 | 38 | 15.8 | 505855 | 90.7 |
| Dry land | 31 | 20.1 | 2136 | 2.2 | 21 | 13.4 | 2162 | 0.3 | 29 | 12.0 | 3388 | 0.6 |
| Dry land with abundant swamp zones | 16 | 10.4 | 633 | 0.7 | 36 | 22.9 | 32488 | 5.2 | 76 | 31.5 | 17739 | 3.2 |
| Total | 154 | 100 | 97250 | 100 | 157 | 100 | 630405 | 100 | 241 | 100 | 558053 | 100 |

* All volumes in barrels.

Source Awobajo 1983

Most oil is spilt in the swamps (mainly the mangroves of the Niger delta) or in offshore waters; however there is considerable variation from year to year in the relative amounts of oil entering these two environmental types.  Offshore slicks may drift into the swamps, where clean-up is particularly difficult.  The statistics of Table 12 underlie two problems which recur in many parts of the world, namely the treatment of offshore slicks before they move inshore (where they may be more difficult to clean up), and the treatment of 'oil trap' areas such as mangrove swamps. (See Table 14).

Table 13. Distribution of oil pollution incidents
in and around U.S. waters 1981

| LOCATION | No. of incidents | % |
|---|---|---|
| **Inland** | | |
| Open internal waters | 226 | 2.6 |
| River channels | 2,055 | 23.3 |
| Ports and harbors | 484 | 5.5 |
| Beaches, non-navigable waters | 802 | 9.1 |
| Territorial sea (shore-3 miles) | 0 | 0.0 |
| Contiguous zone (3-12 miles) | 0 | 0.0 |
| High seas (12 miles or more) | 0 | 0.0 |
| Total | 3,567 | 40.4 |
| | | |
| **Atlantic** | | |
| Open internal waters | 0 | 0.0 |
| River channels | 930 | 10.5 |
| Ports and harbors | 720 | 8.2 |
| Beaches, non-navigable waters | 114 | 1.3 |
| Territorial sea (shore-3 miles) | 202 | 2.3 |
| Contiguous zone (3-12 miles) | 14 | 0.2 |
| High seas (12 miles or more) | 13 | 0.1 |
| Total | 1,993 | 22.6 |
| | | |
| **Pacific** | | |
| Open internal waters | 0 | 0.0 |
| River channels | 81 | 0.9 |
| Ports and harbors | 855 | 9.7 |
| Beaches, non-navigable waters | 66 | 0.7 |
| Territorial sea (shore-3 miles) | 241 | 2.7 |
| Contiguous zone (3-12 miles) | 25 | 0.3 |
| High seas (12 miles or more) | 8 | 0.1 |
| Total | 1,276 | 14.5 |
| | | |
| **Gulf** | | |
| Open internal waters | 0 | 0.0 |
| River channels | 216 | 2.4 |
| Ports and harbors | 350 | 4.0 |
| Beaches, non-navigable waters | 41 | 0.5 |
| Territorial sea (shore-3 miles) | 428 | 4.9 |
| Contiguous zone (3-12 miles) | 392 | 4.4 |
| High seas (12 miles or more) | 262 | 3.0 |
| Total | 1,689 | 19.1 |
| | | |
| **Great Lakes** | | |
| Open internal waters | 21 | 0.2 |
| River channels | 88 | 1.0 |
| Ports and harbors | 69 | 0.8 |
| Beaches, non-navigable waters | 56 | 0.6 |
| Total | 234 | 2.7 |

(continued)

Table 13.   Continued

| Other | | |
|---|---|---|
| Open internal waters | 0 | 0.0 |
| River channels | 2 | 0.0 |
| Ports and harbors | 38 | 0.4 |
| Beaches, non-navigable waters | 5 | 0.1 |
| Territorial sea (shore-3 miles) | 5 | 0.1 |
| Contiguous zone (3-12 miles) | 3 | 0.0 |
| High seas (12 miles or more) | 1 | 0.0 |
| Other | 7 | 0.1 |
| **Total** | **61** | **0.7** |

| Grand totals | | |
|---|---|---|
| Open internal waters | 247 | 2.8 |
| River channels | 3,372 | 38.2 |
| Ports and harbors | 2,516 | 28.5 |
| Beaches, non-navigable waters | 1,084 | 12.3 |
| Territorial sea (shore-3 miles) | 876 | 9.9 |
| Contiguous zone (3-12 miles) | 434 | 4.9 |
| High seas (12 miles or more) | 284 | 3.2 |
| Other | 7 | 0.1 |
| **Grand total** | **8,820** | **100.0** |

Source USCG 1983

The greatest numbers of incidents affect river channels, ports and harbors, beaches and non-navigable waters.  The same conclusion may be drawn from numerous other statistics, for example the 1982 data for U.K. waters published by ACOPS 1983.  In summary, ACOPS reported 204 open sea pollutions and 381 inshore ones (tidal rivers/ estuaries 88, bay/nearshore 42, beach/shore 60 and ports 191).  The Royal Commission on Environmental Pollution (1981) came to the following conclusion about U.K. contingency planning 'our main concern with regard to contingency planning is with the arrangements onshore which we believe have received inadequate attention.  The tendency has been to concentrate attention on action at sea and to regard the clean-up of any oil that reaches the coast as a costly and unpleasant task but not one that calls for much expertise or urgency.  We would not accept this view which we think has encouraged the neglect of possibilities for reducing damage by action on the shore and in inshore waters.'

## VULNERABILITY CLASSIFICATIONS

Vulnerability classifications are a type of environmental classification with built-in guidance on fate and ecological effects of oil, and clean-up approach.  They are most reliably based on extensive case history experience, and for this reason are available mainly for coastal areas.  One of the most widely used systems in contingency planning is that of Gundlach and Hayes 1978 (Table 14). The original system has been modified for particular requirements, e.g. for the Alaskan coast (Gundlach and Hayes 1982). Vulnerability classifications for marine areas other than the coast and for freshwater and terrestrial habitats would be useful, but are less well developed.

Table 14. Environmental classification in order of increasing vulnerability to oil spill damage (modified from Gundlach and Hayes 1978)

| Vulner-ability index | Shoreline type | Comments |
|---|---|---|
| 1 | Exposed rocky headlands | Wave reflection keeps most of the oil offshore. Clean-up frequently unnecessary. |
| 2 | Eroding wave-cut platforms | Wave-swept. Most oil removed by natural processes within weeks. |
| 3 | Fine-grained sand beaches | Oil does not usually penetrate far into the sediment, facilitating mechanical removal if necessary. Otherwise, oil may persist several months. (Recent evidence suggests that water table movements in sediments are a factor affecting degree of penetration). |
| 4 | Coarse-grained sand beaches | Oil may sink and/or be buried rapidly making clean-up difficult. Under moderate to high energy conditions, oil will be removed naturally within months from most of the beachface. |
| 5 | Exposed, compacted tidal flats | Most oil will not adhere to, nor penetrate into the compacted tidal flat. Clean-up is usually unnecessary, except to prevent the oil from going elsewhere. |
| 6 | Mixed sand and gravel beaches | Oil may undergo rapid penetration and burial. Under moderate to low energy conditions, oil may persist for years. |
| 7 | Gravel beaches | Same as above. Clean-up should concentrate on high-tide swash area. A solid asphalt pavement may form under heavy oil accumulations. |
| 8 | Sheltered rocky coasts | Areas of reduced wave action. Oil may persist for many years. Clean-up is not recommended unless oil concentration is very heavy. |
| 9 | Sheltered tidal flats | Areas of low wave energy and high biological productivity. Clean-up is not recommended unless oil accumulation is very heavy. These areas should receive priority protection by using booms or oil sorbent materials. |
| 10 | Saltmarshes and mangroves | Most productive of aquatic environments. Oil may persist for years. Cleaning of salt-marshes by burning, cutting or stripping should be undertaken only if heavily oiled. Mangroves should not be altered. Protection of these environments by booms or sorbent material should receive first priority. |

The implementation of a vulnerability classification or environmental sensitivity index in any particular area requires that existing data (geomorphological, biological and socioeconomic) are collected, and augmented as necessary with new ground surveys and low-level aerial observations. The sensitivities of the various parts of the area under consideration are deduced from these data, and presented on maps, usually using colors and symbols. Spill response information (e.g. access roads, placement sites for booms) is included on the maps, which thus summarize much of the information necessary for effective and cost-efficient spill response.


MANAGEMENT PROBLEMS

Finally, as spill response is rarely perfect, it may be useful to list some response management problems. The main ones seem to be:

1.  Inadequate contingency planning, including inadequate organization resulting from lack of clear lines of authority, and lack of identification of sensitive areas and appropriate clean-up techniques.

2.  Delays pending the allocation of responsibility.

3.  Lack of practice in spill response. Any spill response plan benefits from regular rehearsals which help to pinpoint weaknesses such as communication blocks or problems in handling equipment. Improvements in response can then be implemented before a 'real' spill.

4.  Inadequate communication of existing clean-up knowledge. For example, dispersants may be used in inappropriate circumstances (such as some beach sediments where oil penetration will probably be enhanced by dispersant use) because appropriate existing knowledge has not filtered through to the team actually doing the work. This problem has to be dealt with by thorough initial training and regular updates.

5.  Public pressure for any kind of action. Public concern and media coverage about particular incidents may be high. A government or local authority may feel that it has to be seen to be doing something even if the action is of little practical use. For example, much money and effort may be spent in spraying dispersants on intractable offshore fuel oil because the oil is threatening holiday beaches. The

scientific decision may be that it is best to let the oil
come ashore on a 'sacrificial beach' and then attempt various
mechanical clean-up techniques, but abandonment of offshore
activities may be unacceptable politically.

In some cases the 'vulture effect' (lots of uninvited experts
getting in the way) can be a nuisance for the spill response manager.

CONCLUSIONS

1.  Oil spills are a common type of ecoaccident.  Several thousand
    spills have been at least partly documented.

2.  Classification by source and cause is useful in highlighting
    problem areas (such as the high proportion of tanker spills
    which occur during loading or discharging, or external
    corrosion of pipelines).  It is difficult, however, for any
    scheme to adequately identify and classify the various types
    of human error which are the primary cause of many, probably
    most, spills.  The human error factor may be hidden in a
    heading such as 'hose failure'.

3.  Some causes are specific to particular geographical areas so
    require local engineering and human solutions.

4.  Classification by size shows that the majority of spills are
    small, but that total volume spilt may, for particular periods
    and areas, be dominated by a few major incidents.  This points
    to the need for at least two tiers of clean-up - local teams
    for dealing with small spills and co-operative regional or
    national plans for large ones.

5.  A great variety of clean-up methods are available.  Effective
    contingency planning requires that these be considered
    together with the possible environmental distribution of
    spilt oil in the area under consideration.  Different
    environmental categories require different clean-up approaches,
    as do different oil types.

6.  Clean-up teams require thorough training with regular up-
    dating and rehearsals.

7.  Accidental spills attract much publicity but, compared with
    routine shipping operations and land-based discharges, account
    for only a minor proportion of the oil entering the
    environment.

ACKNOWLEDGEMENTS

I am grateful to Dr Ian White and Mrs Susan Evans for assistance in preparing this paper.


REFERENCES

ACOPS (1983). Survey of marine oil pollution around the U.K. 1982. Advisory Committee on Pollution of the Sea. 24 pp.

Awobajo, S.A. (1983). An analysis of oil spill incidents in Nigeria: 1976-1980. In: The Petroleum Industry and the Nigerian Environment (A.A. Thomopulus, Ed.) 57-63. Thomopulus Environmental Pollution Consultants, Lagos, Nigeria.

Baker, J.M. (1983). Impact of oil pollution on living resources. The Environmentalist 3, Supplement No.4. IUCN Commission on Ecology Papers No.4 48 pp.

CONCAWE (1981). A field guide to coastal oil spill control and clean-up techniques. Report No.9/81. CONCAWE, Den Haag. 112 pp.

CONCAWE (1983). Performance of oil industry cross-country pipelines in Western Europe. Report No.9/83. CONCAWE, Den Haag. 18 pp.

Gundlach, E.R. and Hayes, M.O. (1978). Vulnerability of coastal environments to oil spill impacts. Marine Tech. Society Journal, 12, 18-27.

Gundlach, E.R. and Hayes, M.O. (1982). The oil spill environmental sensitivity index applied to the Alaskan coast. In: Proceedings of the Fifth Arctic Oilspill Program Technical Seminar. June 16-18, 1982, Edmonton. 311-323.

IMCO (1981). Petroleum in the marine environment. Inputs of petroleum hydrocarbons into the ocean due to marine transportation activities. Inter-Governmental Maritime Consultative Organization, November, 1981. Note: IMCO became IMO (International Maritime Organization) after 22nd May, 1982.

Meikle, K.M. (1981). An oil slick igniter for remote areas, In: Proceedings of the 1981 Oil Spill Conference (Prevention, Behaviour, Control, Clean-up). March 2-3 1981, Atlanta. 617-621. API/EPA/USCG.

RCEP (1981). Oil pollution of the sea. Eighth report of the Royal Commission on Environmental Pollution, October, 1981. HMSO, London, 307 pp.

USCG (1983). Polluting incidents in and around U.S. waters. United States Coast Guard Report COMDTINST M16450.2E 72 pp.

Wardley-Smith, J. (Ed.) (1983). The control of oil pollution. Graham and Trotman Ltd, London. 2nd ed. 272 pp.

SCENARIO OF AN ACCIDENT WITH TRICHLOROETHYLENE: AN ATTEMPT AT THE
PREDICTION OF LONG-TERM ECOTOXICOLOGICAL EFFECTS ON THE BASIS OF
PHYSICO-CHEMICAL, TOXICOLOGICAL, AND ECOTOXICOLOGICAL DATA*

Horst W. Peter

Umweltbundesamt
Bismarckplatz 1
D-100 Berlin 33, FRG

INTRODUCTION

This chapter attempts to predict the ecotoxicological potential
of a chemical that enters the environment after an accident, based
on all data available about the substance.  The prediction will be
as general as possible and characteristics of the region will not
be taken into account.  For this prediction, data will be needed on

physical and chemical properties of the substance
its effects on biotic systems
degradability and accumulation of the substance
decomposition-, combustion- and hydrolisis-products toxicity
    for man and animals (short-term and long-term effects)
volume produced
additive and multiplicative effects with other substances

CHEMICALS ACT

New substances require that certain data be provided automat-
ically under the chemical control acts of the EEC member states
and the Toxic Substances Control Act (TSCA) of the United States.
According to "Directive 79/831 of the European Community" all
chemicals must be tested and notifiable prior to being placed on
the market.  Notification to the competent authority means presenting

---

*This article represents the opinion of the author which does not
 necessarily correspond with that of the Umweltbundesamt

extensive, clearly defined information about the substance.  The
amount of information depends on the marketing volume and follows
a step-system.  If the marketing quantity per producer or importer
reaches 1 ton/year, information must be provided mainly about the
identity of the substance, the application form, the estimated
production, precautions for handling, physico-chemical properties,
and toxical- and ecotoxical properties.  These principles are listed
in Annex VII of EEC Directive 79/831 (6th amendment of 67 directive).
The competent national authority can require the notifier to submit
information listed in Annex VIII of the Directive, level 1, when
the production volume is in the range 10 to 1000 tons/year (50 to
5000 tons cumulative).  When a chemical reaches 1000 tons/year or
passes 5000 tons total marketing quantity per notifier, additional
experiments are necessary, which are chosen by the notifier and
the competent authority, who both consider the characteristics of
the substance (Schmidt-Bleek, 1981). The German Chemicals Act is
in the first rank an act for newly produced or imported substances,
and, therefore, it provides only data on chemicals produced and
marketed from September, 1981.  All "old substances," the number
of which is in the range of about 50,000, can only be tested on
the basis of the Chemicals Act if there is a well-founded suspicion
of their hazardous properties.  On the other hand, a great deal of
data can be gathered about "old substances" from the literature.
Following is the procedure for dealing with a known substance as
a chemical to be tested according to our German Chemicals Act.
This procedure also examines if the data required for the Chemicals
Act are sufficient for predicting ecotoxic effects of substances
introduced into the environment after an accident.


DATA ON TRICHLOROETHYLENE

Trichloroethylene is a chlorinated hydrocarbon and was chosen
for this illustration because:

a.  Data on chlorinated hydrocarbons are plentiful in the
    literature.

b.  An appreciable number of spills with chlorinated hydro-
    carbons in the Federal Republic of Germany have been
    described, partially related to grave pollution of drinking
    water (Ministerium fur Ernahrung, Landwirtschaft, Umwelt
    und Forsten, Baden-Wurttemberg, 1982).

c.  Chlorinated hydrocarbons are slow to biodegrade in soil
    and water and, therefore, may be responsible for long-
    term ecotoxic effects.

d.  The production volume of chlorinated hydrocarbons is
    relatively high.

The following summarizes data about trichloroethylene. The items could have been asked of the notifier by national authories on the basis of the Chemicals Act. Data include levels 0 (base set), 1 (10 to 1000 tons/year), and 2 (1000 to 5000 tons/year). I am assuming that data from the literature is incomplete according to the Chemicals Act, nevertheless the information may be sufficient for some levels of the ecotoxic effects of trichloroethylene.

## ANNEX VII (ALMOST COMPLETE)

1.        IDENTITY OF THE SUBSTANCE
1.1       Name
          1.1.2-Trichloroethylene
1.1.1     Names in the IUPAC nomenclature
1.1.2     Other names
          Ethylene trichloride, Trichloroethene, Ethinyl trichloride
1.1.3     CAS number
          79-01-6
1.2       Empirical and structural formula

$$\mathrm{Cl} \diagdown \atop \mathrm{Cl} \diagup \quad C = C \diagup^{\mathrm{Cl}}_{\diagdown \mathrm{H}}$$

1.3       Composition of the substance
1.3.1     Degree of purity (%)
1.3.2     Nature of impurities, including isomers and by-products:
          epoxybutane and epichlorohydrin as stabilizers
1.3.3     Percentage of (significant) main impurities:  about 0.3%
1.3.4     Spectral data (UV, IR, NMR)
1.4       Methods of detection and determination:  mainly gas
          chromatography

2.        INFORMATION ON THE SUBSTANCE
2.1       Proposed uses
2.1.1     Types of use
          Describe:  the function of the substance:
                     Metal degreasing 80%
                     dry cleaning 10%
                     solvent 10%
2.1.2     Fields or application approximate breakdown
2.2       Estimated production and/or imports for each of the
          anticipated uses or fields of application
2.2.1     Overall production and/or imports in order of tons per
          year:  1980 in the FRG: 47,000 tons (Ministerium fur
          Ernahrung, Landwirtschaft, Umwelt und Forsten, Baden-
          Wurttemberg, 1982)
2.3       Recommended methods and precautions concerning:
2.3.1     handling

2.3.2  storage

2.3.3  transport

2.3.4  fire (nature of combustion gases or pryolysis, where proposed uses justify this)

2.3.5  other dangers, particularly chemical reaction with water

2.4  Emergency measures in case of accidental spillage

2.5  Emergency measures in case of injury to persons (e.g., poisoning)


3.  PHYSICO-CHEMICAL PROPERTIES OF THE SUBSTANCE (Atri, 1983)

3.1  Melting point: $-73^{o}C$

3.2  Boiling point:  $87^{o}C$

3.3  Relative density: $1.4642 \; d_4^{20}$

3.4  Vapour pressure: $7.5 \times 10^3 \; Pa$ at $20^{o}C$

3.5  Surface tension:  in water solution $39 \; nMm^{-1}$ at $30^{o}C$

3.6  Water solubility:  $1.1 \; g/l$ at $20^{o}C$
                         $1.25 \; g/l$ at $60^{o}C$

3.7  Fat solubility:  $714 \; mg/100 \; g$ olive oil

3.8  Partition coefficient
     n-octanol/water:    $\log P_{OW} = 2.26$

3.9  Flash point:  $32.0^{o}C$

3.10  Flammability:  $410^{o}C$

3.11  Explosive properties:

3.12  Auto-flammability

3.13  Oxidizing properties


4.  TOXICOLOGICAL STUDIES

4.1  Acute toxicity

4.1.1  Administered orally
     LD50 rat = 4920 mg/kg (NOSH, 1977)
     LD50 mouse = 3000 mg/kg (World Health Organization, 1980)
     Effects observed, including in the organs:

4.1.2  Administered by inhalation
     LC50 rat (1 h) = 26,000 ppm (Vernot et al., 1977)
     Effects observed, including in the organs: histological changes in the liver comprised edema, some ballooning of cells in the central and mid-zonal areas and a few small areas of necrosis

4.1.3  Administered cutaneously:
     LD50 rabbit = 29.3 g/kg (Smyth et al., 1962) (percutaneous)
     Effects observed, including in the organs:

4.1.4  Substances other than gases shall be administered via two routes, at least one of which should be the oral route. The other route will depend on the intended use and on the physical properties of the substance.
     Gases and volatile liquids should be administered by inhalation (a minimum period of administration of 4 h).
     In all cases, observation of the animals should be carried out at least 14 days.

Unless there are contra-indications, the rat is the preferred species for oral and inhalation experiments. Experiments 4.1.1, 4.1.2, and 4.1.3 shall be carried out on both male and female subjects.

4.1.5   Skin irritation
Trichloroethylene produces severe skin irritation when tested neat on the rabbit skin using a 24-h occlusive dressing (Duprat et al., 1976)

4.1.6   Eye irritation
0.1 ml Trichloroethylene applied directly to the eye produces a mild-moderate conjunctivity with some epithelial abrasions (Duprat et al., 1976)

4.1.7   Skin sensitization

4.2     Sub-acute toxicity

4.2.1   Sub-acute toxicity (oral)
0.56-5.62 g/kg (rats) was applied in corn oil by gavage 5 days/week for 6 weeks. Animals were observed for two further weeks.
All animals died following dosage with 5.62 g/kg, but no animals died at 3.1 g/kg or below. Decreasing body weight gain of 20% was noted in females at 1.78 g/kg and above, in males at 3.16 g/kg and above. Other signs of toxicity noted at 3.16 g/kg and above were hunching, urine-stained fur, alepocia, and laboured breathing. Gross examination at autopsy revealed a dilated renal pelvis in one male and a dark red kidney medulla in another at 1.78 g/kg. Incidental findings of large abscessed regions in all lung lobes of two test animals were noted (National Cancer Institute, 1975).

4.2.1   Sub-acute toxicity (inhalation)
Rats and rabbits exposed to 3,000 ppm of trichloroethylene for 7 h/days, 5 days/week during a 6-month period experienced an increase in liver and kidney weight. At lower concentrations (400 ppm), the increase in liver and kidney weight was still noticiable in rats, but much less dramatic than before (Status Assessment of Toxic Chemicals: Trichloroethylene, 1979).

4.3     Other effects

4.3.1   Mutagenicity (including carcinogenic prescreening test)
Trichloroethylene is reported to be mutagenic in the Ames test in Salmonella typhimurium strain TA 100 with a mammalian microsome fraction and in TA 1535 and TA 1538 without mammalian metabolic activation. In the presence of a metabolic activation system, trichloroethylene is reported to induce reverse mutation in Escherichia coli strain K 12 and frameshift and base substitution in Saccharomyces cerevisae XV185-14C (Environmental Research Program of the BMI, 1979).

5.      ECOTOXICOLOGICAL STUDIES

5.1     Effects on organism

5.1.1  Acute toxicity for fish
LC50 <u>Limanda</u> <u>limanda</u> (96 h) = 16 mg/l (Selenka and Bauer, 1977)
LC50 Golden orfe (48 h) = 136 - 200 mg/l (Organization for Economic Cooperation and Development (OECD), 1981)
0.0005 ml trichloroethylene/l applied for 60 h to black mollies led to losses in weight, to changes in the morphology of liver, and to an enhanced mortality (Loekle et al., 1983).

5.1.2  Acute toxicity for <u>Daphnia</u>
LC50 <u>Daphnia</u> <u>magna</u> (48 h) = 65 mg/l (Canton and Adema, 1978)
LC50 <u>Daphnia</u> <u>pulex</u> (48 h) = 45 mg/l (Canton and Adema, 1978)

5.2  Degradation
- biotic:  incubation aerobic (in presence of wastewater microorganism) or anaerobic (in presence of methane bacteria):
ambiguous (Buumer et al., 1981)
biotic and chemical degradation in seawater:
2.5 years (Pearson and McConnell, 1975)
- abiotic:  $t_{1/2}$ in water: 2.6 - 6 years
in air:  6 weeks
(Landesanstalt fur Wasser und Abfall, Nordrhein-Westfalen, 1977)
- photodegradation:  $t_{1/2}$ = 8 days (.OH) = 5 x $10^5 cm^{-3}$
(OECD, 1981)
- degradation products:  mainly HCl, $COCl_2$
(Landesanstalt fur Wasser und Abfall, Nordrhein-Westfalen, 1977).


ANNEX VIII (almost complete)
Additional Information and Tests Required Under Article 6(5)

Level 1: 10 to 1000 tons/year
Taking into account:
- current knowledge of the substance,
- known and planned uses,
- results of tests carried out in the context of the base set


TOXICOLOGICAL STUDIES

Fertility study
Male mice had been exposed to 0.50, 202, or 450 ppm for 24 h, with 50 mice per dose level.  There were no biologically significant effects on pregnancy rates or pre- or post- implantation losses at any dose level -- any variation was within the range of controls (Slacik-Erben et al., 1980).

Teratology study
After exposure to 1.8 ppm trichloroethylene to rats, no signs of
maternal toxicity had been detected (Dorfmuller et al., 1979).

Sub-chronic and/or chronic toxicity study, including special
studies (one species, male and female, most appropriate route of
administration)

Additional mutagenesis studies (including screening for
carcinogenesis)


ECOTOXICOLOGY STUDIES

An algal test: one species, growth inhibition test
LC50 (unicellular algae)        =     8 mg/l (McConnell et al., 1975)
TLC (Microcystis aeruginosa)    =    63 mg/l (Bringmann and Kuhn, 1980)
TLC (Scenedesmus quadricauda) = 1000 mg/l (Bringmann and Kuhn, 1980)

Prolonged toxicity study with Daphnia magna (21 days; this
study should also include determination of the "no-effect-level"
for reproduction and the "no-effect-level" for lethality).

Test on a higher plant

Test on an earthworm

Prolonged toxicity study with fish (e.g. Oryzias, Jordanella,
etc.; at least a period of 14 days; this study should also include
determination of the "threshold level").

Test for species accumulation; one species, preferably fish
(e.g. Poecilla reticulata). Prolonged biodegradation study, if
sufficient (bio)degradation has not been proved by studies laid
down in the Annex VII, another test (dynamic) should be chosen with
lower concentrations and with a different inoculum (e.g. flow-
through system): 2 -18 months in unadapted soils (Selenka and
Bauer, 1977).

LEVEL 2: 1000 to 5000 tons/year

The test program shall cover the following aspects, unless
there are strong reasons to the contrary supported by evidence,
that it should not be followed:
- chronic toxicity study,
- carcinogenicity study,
- fertility study (e.g., three-generation study; only if an effect
  on fertility has been established at Level 1),

- acute and sub-acute toxicity study on second species: only if
  results of biotransformation studies and studies on pharmacokine-
  tics may lead to such studies.

### Chronic toxicity study

Effect of long-term exposure of trichloroethylene on the behavior
of mongolian gerbils (<u>Meriones</u> <u>unguiculatus</u>) (Kjellstrand et al.,
1981). These tests had been done because there is evidence that
trichloroethylene might cause permanent damage to the central nervous
system, provided the time of exposure is long enough. Two groups
of mongolian gerbils were continuously exposed to 150 ppm trichloroethylene for 71 and 106 days, respectively. The behavior of the
animals was tested using a maze, baited with sunflower seeds. Comparison between the trichloroethylene- and air-exposed animals
showed differences in the number of correct choices and the number
of seeds consumed in the maze.

Chronic effects of trichloroethylene on rat behavior (Battig
and Grandjean, 1963): 400 ppm were given for 8 h/day during 5 days/
week for 44 weeks to male rats. No increase in mortality or decrease
in weight gain was noted. Neuromuscular efficiency as measured by
swimming tests was reduced significantly.

### Carcinogenicity study

Trichloroethylene is described as cancerogenic in mice, producing
a significant incidence of hepatocellular carcinoma when administered
orally or inhalative at doses of about 500 - 2000 mg/kg (Henschler
1981; National Cancer Institut, 1975). No significant difference
was observed in rats between the incidence of neoplastic lesions
in trichloroethylene-treated and control animals (Henschler, 1981).

### Toxicokinetic studies

Metabolism at exposition experiments with men (Triebig et al.,
1976; Ertle et al., 1972). (The Chemicals Act does not prescribe
any tests with humans of course.)

Exposition: 5 days per 6 h with 50 or 100 ppm

| | | |
|---|---|---|
| 50 ppm trichloroethanol in the blood: | 0.5 - 2.0 | µg/24 h |
| trichloroethanol in the urine: | 120 - 200 | mg/24 h |
| trichloroacetic acid in the urine: | 20 - 100 | mg/24 h |
| 100 ppm trichloroethanol in the blood: | 3.0 - 5.0 | µg/ml |
| trichloroethanol in the urine: | 190 - 300 | mg/24 h |
| trichloroacetic acid in the urine: | 20 - 250 | mg/24 h |

Exposition: 5 days per 4 h with 50 ppm

| | | |
|---|---|---|
| trichloroethylene in the blood: | 0.4 - 0.8 | µg/ml |
| trichloroethanol in the blood: | 1.0 - 3.0 | µg/ml |
| trichloroethanol in the urine: | 45 - 120 | mg/24 h |
| trichloroacetic acid in the urine: | 3 - 90 | mg/24 h |

Ecotoxicology:
Additional tests for accumulation, degradation, and mobility.  The
purpose of this study should be to determine any accumulation in
the food chain.  For further bioaccumulation studies, special attention should be paid to the solubility of the substance in water and
its n-octanol/water partition coefficient.
The results of the Level 1 accumulation and the physico-chemical
properties may lead to a large-scale flow-through test.

Prolonged toxicity study with fish (including reproduction).
Additional toxicity study (acute and sub-acute) with birds (e.g.,
    quails) if accumulation  factor is >100.
Additional toxicity study with other organisms (if this proves
    necessary).
Absorption-desorption study where the substance is not particularly
    degradable.

    Bacteria test
TLC for "Pseudomonas putida" = 65 mg/l (Bringmann and Kuhn, 1980)


EVALUATION OF DATA

    Data on the molecular weight and structural formula and spectral
data serve to identify the substance.  Information on the proposed
use of the substance is not of interest with regard to an accidental
spill of trichloroethylene.  Declarations about precautions and
emergency measures would have been very important in the case of
an accidental spill so that immediate measures could be taken to
insure the safety of humans and animals.  Of course, no conclusions
can be drawn as to the long-term effects.

Physico-Chemical Data

    Of what importance are the physico-chemical data on the sub-
stance for predicting the course of an accident with trichloroethylene?  Above all, such data allow predictions about the distribution of the chemical between the environmental compartments of
air, water, and soil.  The relative density of trichloroethylene
is 1.46, its solubility in water is 1.1 g/l at 20°C and, therefore,
rather poor.  This information suggests that trichloroethylene
behaves  similarly to oil when spilled into the environment.  This
means that the velocity of the infiltration into the soil is about
twice that of water, and, dependent on the heterogenity of the soil
and the degree of water saturation in the soil, the distribution
of the trichloroethylene may be greater with increasing depth.
Trichloroethylene will stop seeping after reaching the impermeable
zone.  From this point, trichloroethylene can be transported by
groundwater.  Figure 1 shows the possible course trichloroethylene

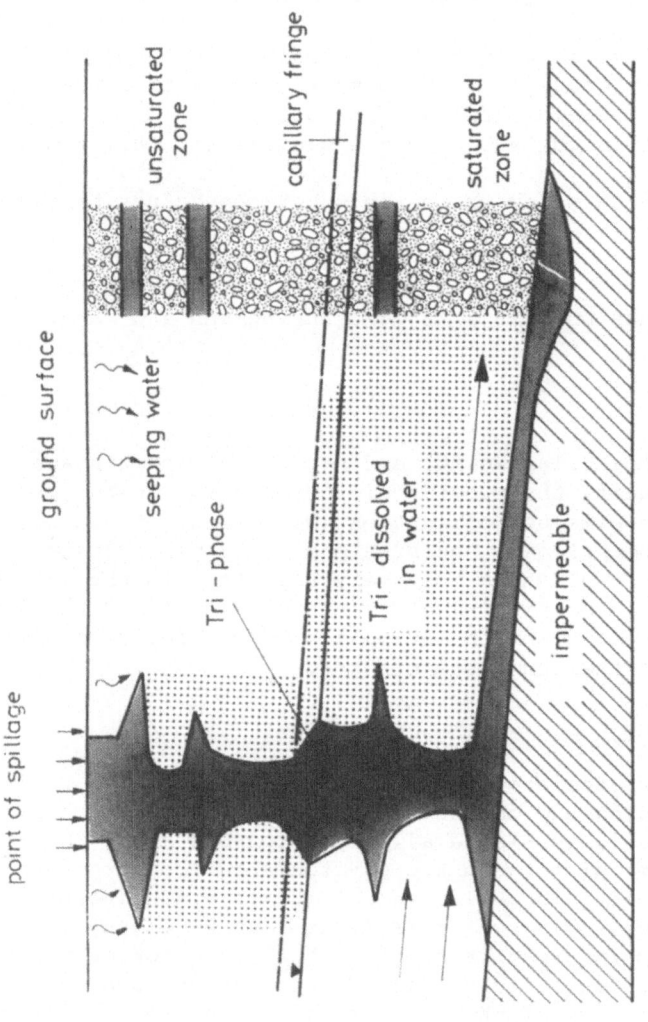

Fig. 1  Trichloroethylene (Tri) migration pattern
(Ministerium für Ernährung, Landwirtschaft, Umwelt und Forsten Baden-
Württemberg, 1982)

may take when seeping into soil after an accidental spill. The region consisting of pure trichloroethylene is surrounded by an area of aquatic solution of trichloroethylene. In the solubilized form, trichloroethylene has great mobility and can be transported easily into other environmental regions. Because of very high vapor-pressure in extremely porous soil, regions can be formed with trichloroethylene vapor.

Data about solubility in water also give some indication of the rate of biotransformation: enhanced water solubility increases the probability of microorganism acclimatization and, thus, enhances biotransformation. So in the case of trichloroethylene, a low biotransformation is expected.

Considering the fat solubility (714 mg/100 g olive oil) and the n-octanol/water partition coefficient (log $P_{OW}$ = 2.28), we can predict that bioaccumulation of trichloroethylene is possible, but not likely.

Using data about vapor-pressure, solubility in water, molecular weight, and partition coefficient of n-octanol/water, we can calculate, according to the model of Mackay (1979), the equilibrium distribution in water, soil, and air:
air:      92% trichloroethylene
water:     3% trichloroethylene
soil:      5% trichloroethylene

The evidence so far indicates these predictions about trichloroethylene based only on physico-chemical data:

   a.  The main portion of trichloroethylene will be found in the air, but considerable amounts will also be in the soil and a portion will be solubilized in water.
   b.  Biotransformation of the spilled trichloroethylene is not very likely.
   c.  Bioaccumulation of trichloroethylene will occur, only to a small degree, if at all.
   d.  Seeping characteristics of spilled trichloroethylene in the soil will be similar to those of oil.

## Toxicological Data

A great deal of the toxicological data on trichloroethylene is based on the same tests; some of the representative ones were chosen for this discussion. Data on acute oral toxicity and acute inhalation toxicity show that trichloroethylene is slightly toxic. Experiments on sub-acute toxicity demonstrate that trichloroethylene in concentrations of 400 mg/kg or higher causes an increase in liver- and kidney-weight in rats and rabbits. Furthermore, many of the test animals showed less growth.

The mutagenicity of trichloroethylene (or the stabilizers) has been investigated in a number of short-term tests. Positive results were obtained when trichloroethylene was tested using certain micro-organisms, indicating that this compound can produce point mutations. A number of long-term animal studies have been carried out to investigate the carcinogenicity of trichloroethylene, using both the inhalation as well as the oral route. Trichloroethylene has been shown to be carcinogenic in mice when given orally or inhalative at high levels (870 mg/kg and above). Hepatocellular carcinoma was produced. However, it cannot be completely excluded that the stabilizers epoxybutane and epichlorohydrine may be responsible for the carcinogenic potential.

No evidence of any teratogenic effects has been noted. Trichloroethylene has an influence on the central nervous system. The principal effect after acute exposure is depression of the CNS, which causes changes in the behavior of animals.

The value of the acute toxicity data to the prediction of exotoxicological effects is rather limited. Spilled trichloroethylene in the pure form, solubilized in water, or in high air concentrations may kill rodents, and, therefore, other animals living in the ecosystem also.

The ecotoxical consequences of the mutagenic potential of a substance are not clear. However, there is a probability that induction of mutagenicity can lead to the elimination of an animal species from the ecosystem, possibly causing a breakdown of the whole ecosystem. In this context, it may be interesting to note that the Umweltbundesamt is going to inititate a project (Nr. 106 03 037) about the relevance of mutagenicity on the stability of ecosystems.

Sub-acute and long-term toxicity studies have more relevance in the prediction of ecotoxicity than the acute toxicity test. Animals not being affected acutely may suffer an influence when being exposed for a long time. As long-term exposure of trichloroethylene affects the central nervous system, normal behavior in animals deteriorates and their ability to survive in the ecosystem is reduced. It is not clear if the findings on cancerogenic potential of trichloroethylene for mice is valid for other animals. More investigations are needed before coming to a definite conclusion.

The prediction of an ecoaccident using the toxicity data of the base set, Level 1, and Level 2 of the Chemicals Act shows:

a. Spilled trichloroethylene in saturated aquatic solution, in gas or liquid form is able to kill rodents and probably other small animals in a few hours.

128

b.  Trichloroethylene affects animals when exposed for a long time: the main effects observed are changes in the behavior and, in mice, hepatocellular carcinoma.  These effects can lead to the elimination of animal-species from the ecosystem.

c.  Elimination of animal-species can occur also because of the mutagenic potential of trichloroethylene.


## Ecotoxicological Data

Without doubt, ecotoxicological data are the most important data for the prediction of environmental effects of chemicals.  For the base set of Germany's Chemicals Act, only tests for the acute toxicity for fish and Daphnia, as well as tests about the degradability of the substance, are required.

The half-life of trichloroethylene in water is 2.6 - 6 years, and in soil 2 - 18 months.  That means that trichloroethylene belongs to the substances that are poorly degradable in water and soil.  According to the half-life values for abiotic degradation in air (6 weeks) and photodegradation (8 days), it can be assumed that trichloroethylene will not remain for a long time in the air.  However, its lingering period in the air will be sufficient to allow trichloroethylene to enter the soil, rivers, and seas via rainfall.

The LC50 values for fish are in the range of 10 - 200 mg/l and the LC50 values for Daphnia about 50 mg/l, indicating that trichloroethylene is moderately toxic for the fish investigated and for Daphnia.  Unfortunately, data was not available on prolonged toxicity studies with fish required for Level 1 of the Chemicals Act, which would have been important for long-term prediction.  On the other hand, the observations on changes in the morphology of liver in the black molly in an acute toxicity test for Level 0 give an indication that trichloroethylene also has long-term effects on fish.  For Level 1 of Germany's Chemicals Act, furthermore, growth inhibition test for algae, prolonged toxicity studies with Daphnia magna, tests on a higher plant, and tests on an earthworm are required.  In the literature, I only found inhibition tests for algae: for a unicellular algae, a LC50 value of 8 mg/l was described; for Microsystis aeruginosa, a TLC value of 63 mg/l was found; and, for Scenedesmus quadricauda, a TLC value of  1000 mg/l was found.  As algae can be seen as a precursor of higher plants, one can expect pronounced toxic effects of the spilled trichloroethylene on higher plants also.  In an aquatic ecosystem "Microcystis aeruginosa" has important functions in the conversion of the nitrogen of the air into ammonium-ions, the latter being the nitrogenium source of many plants.  Therefore, growth inhibition of "Microcystis aeruginosa" and other algae would represent an interference in the balance of an ecosystem with aquatic fractions. Data on ecotoxicological tests that could have been asked for Level 2 are very rare in the literature, and so I found only bacteria tests: 81 $\pm$ 46 mg/l trichloroethylene (found in a ring test) is able to inhibit the growth of different bacteria. As a consequence of the influence of trichloroethylene on bacteria, we can expect disturbance of processes related with the conversion of N, P-, and C-compounds in the soil, which can also lead to a breakdown of the ecosystem.

In summary, on the basis of the ectoxicological data available, one can make the following predictions for the environmental effects of trichloroethylene:

  a. Spilled trichloroethylene will persist in water and soil
     for an extended period.
  b. Trichloroethylene is acutely toxic for fish and other water
     organisms in concentrations much below the saturation con-
     centration. Long-term effects of trichloroethylene on
     aquatic organisms are probable.
  c. Spilled trichloroethylene inhibits the growth of algae
     and bacteria and, therefore, is able to disturb the balance
     of different ecosystems.

DISCUSSION

Predicting the consequence of a disturbance to ecosystem balance is of great interest. Theoretically, this information should come from the producer of the substance when the production volume reaches Level 2. However, at present, there is still a lack of test systems for such ecotoxicological investigation. Our aim, therefore, is to initiate projects about the development of test systems, especially for Level 2 of our Chemicals Act. One of these projects, for instance, is the project of the Umweltbundesamt Nr. 106 04/018, which deals with the construction of artificial field ecosystems and the influence of chemicals on these systems. The individual artificial ecosystems shall be representative for special regions of nearly all parts of the Federal Republic of Germany. The aim of this project is also to find plants and animals, which are characteristic for the ecosystem concerned, that have essential functions in stabilizing the ecosystem and that are sensitive to chemicals. Therefore we will be able to substitute test organisms that were chosen by chance and that are by no means representative of the ecosystem.

Perhaps it will be possible in the future to use closed model ecosystems (in boxes with soil, plants, animals, water, and the organism known to be representative for the concerned ecosystem) as test systems for predicting the hazardous effects of chemicals (Umweltbundesamt project Nr. 106 04/033).

# REFERENCES

Atri, R. F., 1983, Studie zur Bewertung der Toxizitat wasser-
    gefahrdender Stoffe. Band VI. Trichlorethylen, Forschungsber.
    Umweltbundesamt Nr. 106 07 047.
Battig, K., and Grandjean, E., 1963, Chronic effects of trichloro-
    ethylene on rat behaviour, Arch. Environ. Health, 7:694-699.
Baumer, E.J., Rittmann, B. E., and McCarty, P. L., 1981, Anaerobic
    degradation of halogenated 1- and 2-carbon organic compounds,
    Environ. Sci. Technol., 15(5):596-599.
Bringmann, G., and Kuhn, R., 1980, Comparison of the toxicity
    thresholds of water pollutants to bacteria, algae and protozoa
    in the cell multiplication inhibition test, Water Res., 14:231-
    241.
Canton, J. H., and Adema, D. M. M., 1978, Reproducibility of short-
    term and reproduction toxicity experiments with Daphnia magna
    and comparison of the sensitivity of Daphnia magna with Daphnia
    pulex and Daphnia cucullata in short-term experiments,
    Hydrobiologia, 59:135-140.
Dorfmuller, M. A., Hennes, S. P., York, R. G., and Bornschein, R.L.,
    1979, Evaluation of teratogenicity and behavioural toxicity
    with inhalation exposure of maternal rats to trichloroethylene,
    Toxicology, 14:153-166.
Duprat, P., Delsout, L., and Gradiski, D., 1976, Irritant potency
    of the principal chlorinated aliphatic solvents on the skin
    and ocular mucous membrances of rabbits, Eur. J. Toxicol.,
    3:171-177.
Environmental Research Program of the Federal Minister of the
    Interior (BMI), September, 1979, Research Plan No. 104 01 073.
    Expertise on the Environmental Compatibility Testing of Selected
    Products of the Chemical Industry. Volume 1 - 4, SRI. A.
    Research Contract by Umweltbundesamt.
Ertle, T., Henschler, D., Muller, G., and Spassovski, M., 1972,
    Metabolism of trichloroethylene in man, I: The significance
    of trichloroethylene in long-term exposure conditions. Anh.
    Toxicol., 29:171-188.
Henschler, D., 1981, Prufung von Trichlorathylen auf carcinogene
    Wirkung bei inhalatorischer Aufnahme und Mechanismen der Bio-
    aktivierung im Stoffwechsel. Bundesministerium fur Forschung
    und Technologie. Forschungsbericht HA 82 - 007. Humanisierung
    des Arbeitslebens. Institut fur Toxikologie der Universitat
    Wurzburg.
Kjellstrand, P., Bjerkemo, M., Mortensen, J., Mansson, L., Lanke,
    J., and Holmquist, B., 1981, Effects of long-term exposure
    to trichloroethylene on the behaviour of mongolian gerbils
    (Meriones unguiculatus), J. Toxicol. Environ. Health, 8:787-793.
Landesanstalt fur Wasser und Abfall Nrodrhein-Westfalen, 1977,
    Checkliste zur Beurteilung der Wassergefahrlichkeit von Stoffen,
    KFW-Mitteilungen, 3:1-4.

Loekle, D. M., Schecter, A. J., and Christian, J. J., 1983, Effect
of chloroform, tetrachloroethylene, and, trichloroethylene on
survival, growth, and liver of Poecilia sphenops, Bull. Environ.
Contam. Toxicol., 30:199-205.

Mackay, D., 1979, Finding fugacity feasible, Environ. Sci. Technol.,
13:1218.

McConnell, G., Ferguson, D. M., and Pearson, C. R., 1975, Chlorinated hydrocarbons in the environment, Endeavour (London),
34:13-18.

Ministerium fur Ernahrung, Landwirtschaft, Umwelt und Forsten,
Baden-Wurttemberg, 1982, Verhalten vol leichtfluctigen
Chlorkohlenwasserstoffen im Untergrund und Sanierungsmoglich-
keiten vol Schadensfallen. Informationsschrift.

National Cancer Institut, 1975, "Memorandum of Alert Carcinogenesis
Bioassay of Trichloroethylene," U.S. Department of Health,
Education, and Welfare, Public Health Services, Bethesda,
Maryland.

NIOSH, 1979, Registry of Toxic Effects of Chemical Substances.

Organization for Economic Cooperation and Development (OECD), 1981,
Collection of Minimum Pre-Marketing Sets of Data Including
Environmental Residue Data on Existing Chemicals, OECD Hazard
Assessment Project. OECD Working Party on Exposure Analysis.
Expo-80. 12b/3. Revision. Prepared by Umweltbundesamt,
Berlin.

Pearson, C. R., and McConnell, 1975, Chlorinated $C_1$ and $C_2$ hydrocarbons in the marine environment, Proc. R. Soc. London, B,
189:305-332.

Schmidt-Bleek, F., 1981, Regulated testing of new substances for
environmental exposure and toxicity in the European community,
Regul. Toxicol. Pharmacol., 1:123-146.

Selenka, F., and Bauer, U., 1977, Erhebung von Grundlagen zur
Bewertung von Organochlorverbindungen im Wasser. Abschlußbericht.
Institut fur Hygiene, Ruhr-Universitat Bochum.

Slacik-Erben, R., Rall, R., Franke, G., and Uehleke, H., 1980,
Trichloroethylene vapours do not produce dominant lethal
mutations in male mice, Arch. Toxicol., 45(1):37-44.

Smyth, H. F., Carpenter, C. P., Weil, C. S., Pozzari, U. C.,
Striegel, J. A., 1962, Range-finding toxicity data: List VI.
Am. Ind. Hyg. Assoc. J., 23:95-107.

Status Assessment of Toxic Chemicals: Trichloroethylene, 1979,
Monsanto Research Corp., Dayton, Ohio.

Triebig, G., Essing, H.-G., Schaller, K.-H., and Valentin, H.,
1976, Biochemische und psychologische Untersuchungen an
Trichlorethylenexponierten Probanden; Z bl. Bakt. Hyg. I.
Abt. Orig. B., 163:383-416.

Vernot, E. H., MacEwen, J. D., Haun, C. C., Kinkead, E. R., 1977,
Acute toxicity and skin corrosion data for some organic and
inorganic compounds and aqueous solutions, Toxicol. Appl.
Pharmacol., 42:417-23.

World Health Organization (WHO, EURO), 1980, "Priority Problems
in Toxic Chemicals in Europe."

# A PHILOSOPHY OF THE PROSPECTIVE TESTING OF ENVIRONMENTAL CHEMICALS IN RELATION WITH THE ANALYSIS OF ECOACCIDENTS

H.J. Hueck

Division of Technology for Society TNO

Delft, The Netherlands

## INTRODUCTION

Many regulations exist for the prevention or control of waste problems. All these regulations contain a definition of the waste and/or its properties based either on the nature of the waste itself, or on its effect (consumption of oxygen, for example). The problems are associated especially with noxious wastes have given rise to the special branch of toxicology usually called ecotoxicology although environmental toxicology would perhaps be more appropriate.

In principle ecotoxicology is a very simple branch of science, which can be visualized by the following diagram.

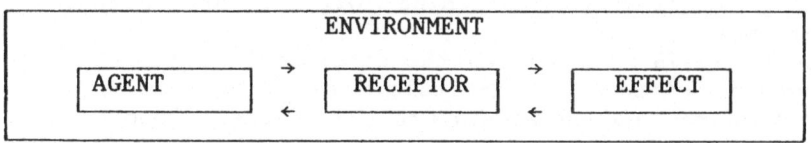

Its aim is to find cause-effect relationships in this system. Awareness of problems usually starts with the observation of (untoward) effects in the environment. Retrospective reasoning may allow the cause to be attributed to a certain agent. However, the complexity of the environment, and the presence of numerous receptors and agents at the same time generally prevent the establishment of an unambiguous cause-effect relationship. In this context it is appropriate to recall the time honoured Kochian principles (Koch, 1882) which have proven to be of great value in a similar situation in the investigation of infectious diseases.

These principles may be easily adapted to the present problem of establishing a straightforward cause-effect relationship in environmental situations:

1. A suspected agent must be found in contact with a vulnerable receptor in such an amount (dose), in such a distribution and under such circumstances that its presence may explain the observed phenomena.
2. The suspected agent should be separated from other similar agents, either by analysis or purification and be kept in a unpolluted and definable state.
3. It must be possible to reproduce with the isolated and defined agent, under controlled conditions, phenomena comparable to the ones originally observed.

It is for the latter purpose that numerous laboratory methods and controlled (semi) field experiments have been developed. Moreover, the availability of such methods provides a possibility to predict to some extent environmental effects of agents wich may potentially be released into the environment. A review of such prospective methods in the context of a book on "ecoaccidents" is appropriate for two reasons viz.

1. A comparison of the effects in the controlled experiment with the real life phenomena in the ecoaccident allows the predictive value of the laboratory method to be checked, and thus its adequacy to be validated.
2. A comparison of the effects in the laboratory experiment with the real life phenomena alo checks the supposed attribution of certain effects to one or more of the agents that are usually present at the same time in "ecoaccidents".

The following chapters will give an outline of the guiding principles in the choice of such methods, and especially of those methods aiming at finding effects on biotic systems other than man. Actual methods exist in such great variety, that an exhaustive treatment in the present context is impossible. Reviews are given in the following multi author publications:

1. Anonymous (1980) "Degradability, Ecotoxicity and Bioaccumulation", Government Publishing Office, The Hague, The Netherlands.
2. Anonymous (1979) "Report on the assessment of potential environmental effects of chemicals; the effects on organisms other than man and on ecosystems". 3 volumes. Report to the Organisation for Economic Cooperation and Development (OECD). Environment Committee - Chemicals group - Chemicals Testing Programme. Paris, France.
3. Anonymous (1981) "OECD Guidelines for Testing Chemicals". OECD 2 rue André Pascal, 75775 Paris CEDEX 16, France.

These reports contain, among other matter, the methods adopted by the EEC for application in the European Community for the imple-

mentation of the so called 6th Amendment relating to the regulation of dangerous substances (Official Journal of the European
Communities 19, No. C260 (4), page 270-285 (5 November 1976).
The methods in these reports are functional ones in that, for regulatory purposes, they try to define chemicals by their properties. Another approach in controlling environmental chemicals is
that by which chemicals or groups of chemicals are placed in so
called black lists, grey lists etc.; the Oslo Convention (Convention for the Prevention of Marine Pollution by Dumping from Ships
and Aircraft, Oslo 15 February 1972, ref. Traktatenblad, 6 June
1972, The Hague, The Netherlands) is an example of this approach.
In regulations of this sort the listed chemicals are included because of known effects. The principles of investigation of these
properties, as referred to above for example in the "OECD" Guidelines", are thus, also implicitly at the basis of this sort of
regulation. The following discussion, will therefore be restricted
to a discussion of principles of these methods. Moreover,
for reasons of space, the discussion will be mainly confined to
methods involving chemicals in the aquous environment


CHOICE OF METHODS AND THEIR PROBLEMS

Introduction

Toxicology is that part of science in which the relationship
between quantities and qualities of a chemical and its effect on
biotic systems is studied. The ecotoxicological approach can be
distinguished from the more traditional toxicological studies by
the nature and identity of the biotic system to be protected. Human toxicology is concerned with the health and safety of each
individual whereas ecotoxicology is concerned with effects on
ecosystems, i.e. on populations and communities.
Another difference which exists between human toxicological testing and ecotoxological testing is the method of administration of
the test compound. In the first the compound is administered directly either by injection or in the food or in the air in inhalation chambers. In ecotoxicological testing the compound is
normally administered in the test medium, frequently water but
also soil or air. In these media, particularly water and soil,
the compound may be modified by degradation, transformation or
adsorption processes, whereby its effect on the test organisms
may be changed, especially in long term testing. Figure 1 illustrates these differences.
Damage to the biotic part of an ecosystem may affect both
function and structure of the system. Function includes the complex relationships through which the ecosystem is mainained. This
is only possible, however, when the material basis for this, viz.

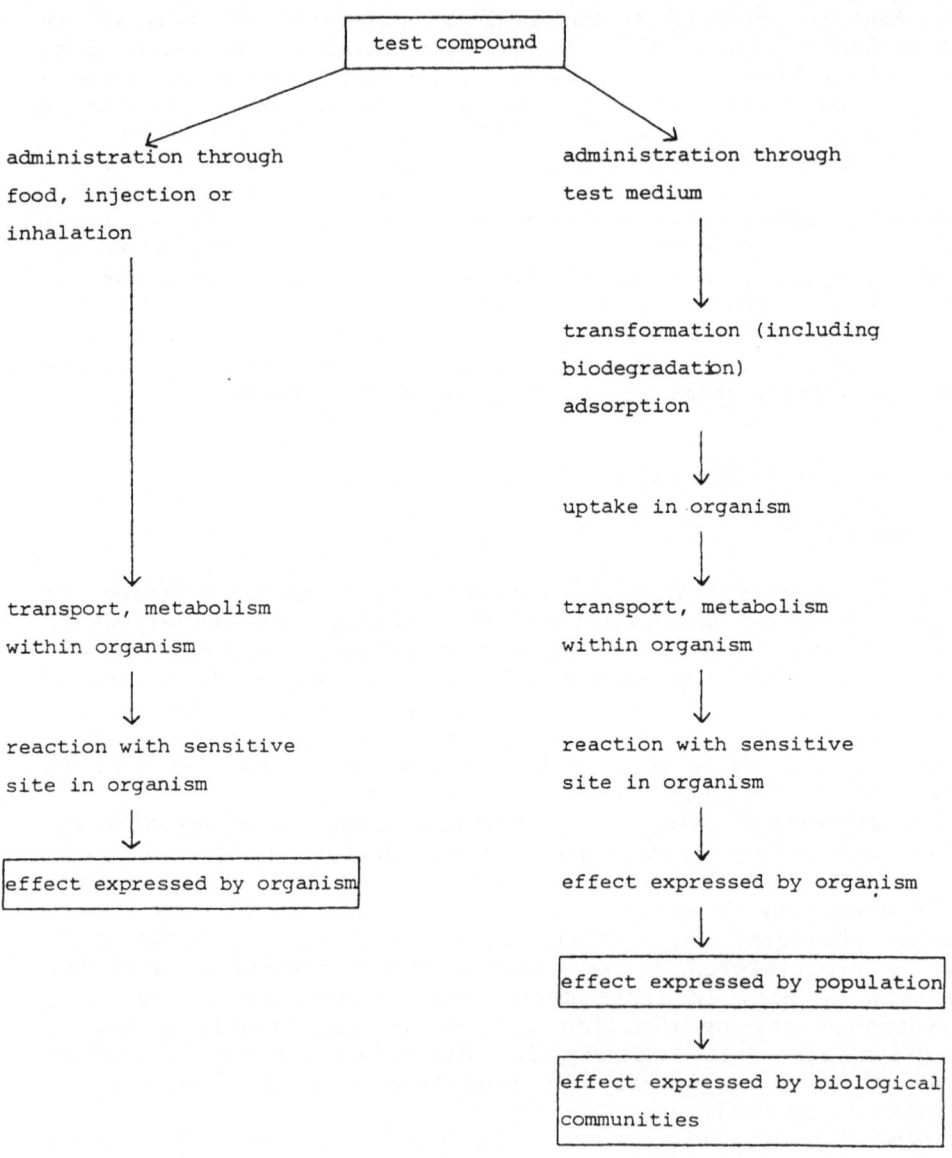

Figure 1: A comparison of human- and ecotoxicological testing.

the proper structure, is present. Structure and function cannot exist without each other, though we may study them separately. In the functioning of ecosystems, trophic relations, involving energy and nutrient transfers, which are part of natural cycles, are of prime importance.

In this respect, the ecologically functional distinction between primary producers, secondary producers, consumers and decay organisms is useful in visualizing important biotic elements of these natural cycles. Environmental disturbances will generally manifest themselves as changes in relative numbers or as changes in the biomass of elements of the system or as both. We may distinguish biodepressive effects (mortality, decrease of growth and reproduction, emigration) and biostimulatory effects (eutrophication, immigration, etc.). Although the latter are not toxic effects for single species, they may be harmful to the ecosystem and as such undesirable.

The chemical under consideration, will be transported to the receptor (the biotic system) via water, air or soil. In this respect the frequently made (and practical) distinction between water, air and soil pollution is not of ecotoxicological significance, the identity of the medium of transport being of minor importance. In assessing the potential effects of chemicals in practice, however, this distinction is of paramount importance, as different methods are used for determining effects in each of these media. To these transport media "food webs" (including the simplified form "food chain") can be added as a separate entity, because of specific phenomena, such as biomagnification and the formation of more or less toxic metabolites, which do not occur in the other types of transfer (fig. 2).

For the assessment of environmental effects, ideally the whole ecosystem should be considered; for pragmatic reasons, however, we are in the laboratory restricted to the analysis of the effects on subsystems. It is only in "Ecoaccidents" that we may observe the whole system.

One of the main difficulties in ecotoxicology is deciding whether

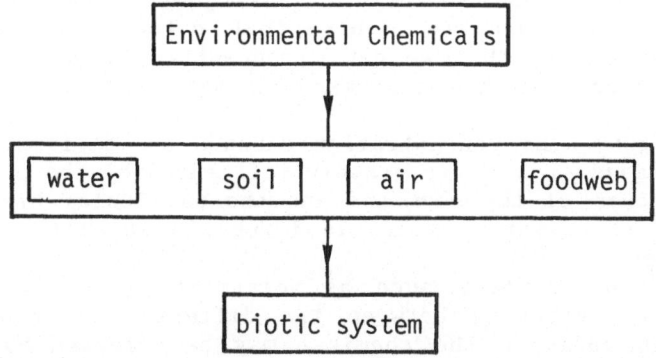

Figure 2: The pathway of environmental chemicals.

or not some observed change is ecologically significant. This problem is made more difficult because of the abilities of biotic systems to repair damage and to adapt themselves to adverse conditions. Ecologically irreversible changes are always to be considered undesirable effects of pollution but reversible effects may, however, also be objectionable, although they may be tolerated under certain circumstances.

It may be assumed that any impairment of the functioning of a biotic system leading to structural damage, especially at the level of communities should be considered as an undesirable pollution effect. From this point of view persistent and/or irreversible effects, are of primary importance in ecotoxicology. It should be borne in mind that some irreversible effects, which are harmful for individual organisms, may be less important at the level of the community (malformations, monsters). The ecological importance of an effect may differ widely between species, e.g. because of differences in reproduction rate or life cycle span.

Important effects include, for example, the main biological events of germination and birth, maturation, reproduction and death which ultimately determine the biomass. Other effects, such as impairment of respiration, increase in mixed function oxidases, changes in pigmentation, behavioural changes, etc., can be considered as secondary effects that may indirectly influence ecologically important parameters. As stated above, effects on biotic systems may be either biostimulatory or biodepressive. The former, of which eutrophication is an example, may be just as serious for the biotic system under consideration as the latter.


## "Dose" in ecotoxicology

In human toxicological testing the word "dose" has a definite meaning, viz. the amount of a compound administered to the test organism, either by injection or through its food. The dose is usually expressed as mg/kg body weight.

In ecotoxicological testing the test compounds are generally not admininstered directly to the test organism but in the test medium. The effects observed are then related to the concentrations in the test medium, so that we get a concentration-effect relationship. Concentrations are expressed in mg/l for water or in mg/kg for soil.

The actual environmental concentration depends in the first place on the amount added, but it is greatly influenced by the physicochemical properties of the chemical and the particular transport medium involved (air, water, soil, food web) in which the biotic system is exposed.

Physico-chemical properties, such as volatility and solubility, determine the distribution between the different environmental compartments. In addition the chemical may be effected by environmental factors before it reaches the living system to be affected. Both physico-chemical stability (resistence to oxidation, hydrolysis, radiation, etc.), adsorption and biodegradability will determine what fraction of the initial environmental concentration will be active in the immediate vicinity of this final living system.

Finally, the dose to which an organism at risk in a system may be exposed is influenced in two ways by processess of bioaccumulation:

- the internal concentrations of chemicals in aquatic organisms are seldom the same as the external concentrations, and substances which are readily absorbed but only slowly lost or released can build up to very high concentrations;
- chemicals may also reach the receptor organisms via food chains or food webs; this is typically the major route for terrestrial animals; it also can lead to enhanced exposure levels at sensitive critical sites for chemicals rapidly absorbed but only slowly lost.

From the above it will be clear that certain key information on physico-chemical properties and the persistence of the chemical under investigation is desirable for the proper planning of ecotoxicological testing, not only because of problems of interpretation of concentration-effect relationships but also for sound operational reasons. For example, different test systems will be necessary for the testing of volatile or hydrophobic compounds. It is furthermore important to check biodegradability before (aquatic) toxicity tests are carried out, since biodegradation not only influences the concentration actually present, but may also lead to erroneous results, because the oxygen depletion may kill the organisms while the compound is not or only slightly toxic.

## Biotic systems and their response in ecotoxicology

Biotic systems may be classified according to their levels of organization. Each level shows specific reactions to pollution; some examples of these responses are given in Table 1.

Traditionally, the organism has received the most attention. In order to establish a link with the supra-organismal levels of the ecosystem i.e. populations and ecosystems Hueck (1978) pointed out that the choice of test organisms should preferably reflect the structure and functioning of the ecosystem under consideration. In this respect the maintenance of the various energy and nutrient cycles (cf. Fig. 3) is important, so that it would appear to be wise to take representatives of the different trophic levels (Lindemann, 1942) into consideration, viz.:

Table 1.  Some effects of pollutants on different levels of biotic organization.

| level of organization | examples of effects | |
|---|---|---|
| | structural aspects | functional aspects |
| biomolecules | DNA defects | impairment of repair and replication; increased mutation rate |
| cells | chromosome defects | impairment of cell metabolism |
| tissues | histopathology, tumor formation | functional impairment |
| organs | pathological morphology, teratogenicity | |
| organisms | pathological morphology, dwarfism | impairment of growth; mortality |
| populations | sex ratio; age dis-stribution | impairment of repro-duction |
| communities | loss of diversity | loss of stability |

Table 2.  Number of species (adapted from Anonymous, OECD, 1979).

| Protista | |
|---|---|
| Bacteria | 1.600 |
| Other | 19.000 |

| Plants | |
|---|---|
| Algae | 15.000 |
| Bryophytes | 24.000 |
| Fungi | 75.000 |
| Vascular plants | 212.000 |

| Invertebrates | |
|---|---|
| "Worms" | 24.000 |
| Mollusks | 70.000 |
| Arthropods | 750.000 |
| Other | 23.000 |

| Vertebrates | |
|---|---|
| Amphibians | 2.800 |
| Reptiles | 7.800 |
| Fish | 20.600 |
| Birds | 8.600 |
| Mammals | 5.000 |

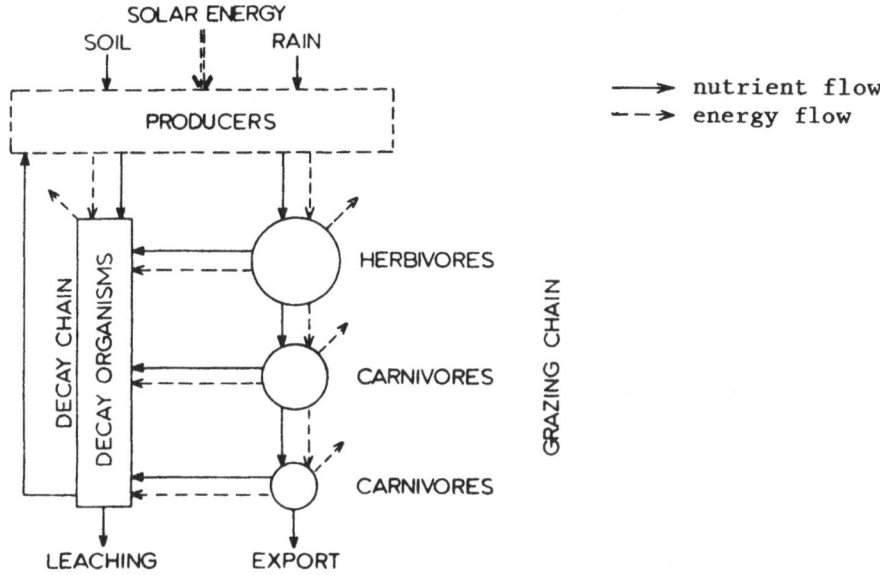

Figure 3: Nutrient and energy flow in an ecosystem (from
Woodwell, 1970).

-   primary production
-   secondary production
-   carnivores
-   decay-organisms

This notwithstanding, the choice of specific tests, with specific
biotic systems in specific cases remains extremely difficult because of the overwhelming number of possiblities. Table 2 gives
a rough indication of the number of species known at the present
moment, from which a choice must be made. Table 3 details the
types of test available. The fact that only few organisms can be
handled in the laboratory limits the possibilities in laboratory
experiments, and the choice of systems in field-experiments may
also be restricted for pragmatic reasons. No single test may
therefore be considered to be representative for an actual situation of pollution. Only with a battery of tests may we expect
to approach the truth as revealed in actual cases of pollution.
Such a battery of test, is advocated by the OECD Guidelines referred to above. A short description of this battery system is
given in the following chapter.

**Table 3.** Types of test used in prospective testing of environmental effects

| level of organization | environmental compartment | | |
|---|---|---|---|
| | air | water | soil |
| cells | - test on enzyme inhibition and enzymatic synthesis<br>- in vitro tests with erythrocytes and chloroplasts<br>- cell respiration and photosynthesis | - tests on enzyme inhibition<br>- cell respiration<br>- heterotrophic activity | - "soil respiration"<br>- heterotrophic activity |
| tissues | - tissue culture tests, a.o. pulmonary tissue<br>- histopathological studies<br>- lung function tests<br>- liver function tests<br>- plant leaf respiration<br>- growth of Avena coleoptiles<br>- pathological morphology of lungs<br>- eye irritation tests with mammals | - tissue culture tests (fish)<br>- rarely histophatological studies<br>- liver function tests (fish)<br>- brain activity and brain damage (fish)<br>- thallus growth of macroalgae | |
| organisms | - exposure chamber tests with higher plants (also seedlings) and lichens<br>- inhalation tests with mammals and birds<br>- rarely, tests with lower animals<br>- tests with bryophytes, criteria: pathological anatomy, growth, respiration, photosynthesis | - test with organisms such as Mollusca, Annelida, Polychaeta, Crustaceae, Echinodermata, Pisces, majority of tests is with fish, criteria: mortality and occasionally growth or reproduction<br>- rarely, tests with hydrozoa and macroalgae | - pot experiments with higher plants<br>- tests with organisms such as Collembola and Lumbricidae, criteria: mortality, growth<br>- rarely, tests with other animals such as: microarthropods and nematodes |
| populations | - occasional experiments with bacteria and algae | - reproduction tests with microorganisms (bacteria, fungi, microalgae, protozoa) or with microalgae, crustaceae and fish | - tests with bacteria, soil algae and protozoa |
| communities | - microcosms, e.g., moss carpets, lichen vegetation, turfgrass | - microcosms with zoo- and phytoplankton, fish<br>- experiments with pond communities | - "litter-bag" (gross heterotrophic decay)<br>- microcosms with soil communities<br>- pot experiments with mixed vegetation |

## The OECD-system of tests

In general the "OECD Guidelines" are designed for use in a stepwise approach to the evaluation of relevant properties of chemicals. In general, 3 levels of testing may be distinguished.
1.  Basic level (level 1)
    At this level simple tests should be used indicating possible effects on a few functionally important types of organisms. Ecological function is considered to be of greater importance than any suggested "representativeness" of particular species for particular environmental compartments.

It is essential that prior to ecotoxicological tests some data are available on physical chemical properties and on the degradability of the compounds to be tested.

2. Comfirmatory level (level 2)

At this level tests should be used which yield more precise information if suspicion as to the acceptability of a chemical has been previously raised at the basis level. Suspicion need not necessarily be based only on the results of ecotoxicological tests. All of the information gathered at the basic level should be used to decide whether or not further testing is necessary and if so, what types of testing should be involved. Such information would include the structure, use and properties of the chemical and potential for dispersion in the environment. Because of the multitude of factors that may give rise to further testing delineation of a single rigorous testing scheme is not possible at this level. A series of tests should be available from which a careful choice can be made.

In general, these tests may be ones of a more complicated nature, related to specific or additional environmental compartments. Also chronic exposure and specific routes of contamination should be considered depending upon the chemical properties and the degree of environmental contamination expected.

3. Definite level (level 3)

In some special cases, e.g. where appreciable environmental concentrations of the chemical are likely to be involved, and/or some indication of possible environmental hazard exists, it may be necessary to assess the effects in experimental systems more closely approaching something like natural conditions, especially with regard to interspecific relations and the functioning of multi-species systems. At present suitable multi-species tests are not readily available and development of these has been identified as a subject area meriting further research. This level of testing of the ecotoxicological assessment of industrial chemicals can be compared with the field testing of the environmental hazards presented by new pesticides. Only general indications of suitable test systems can be given as much depends on the specific properties and use pattern of the chemical under investigation as to what test is selected. The particular test programme necessary must in general be an arbitrary but well informed choice by experts.

The actual tests involved in these 3 levels of test are still under discussion but the following tests are described for the basic level of testing.

1. fish LC50 96 hr: the test protocol should allow for other responses in addition to mortality to be recorded; the spe-

cies selected for testing should be appropriate to the known patterns of use and routes of disposal of the chemical.

2.  Daphnia LC50 14 days: the test is considered to be the most economical test that allows the effect of the chemical on life cycle of an aquatic animal to be evaluated.

3.  Algae LC50 4d (growth inhibition); this test allows a preliminary assessment of the effect of the chemical on the ecosystem function involving carbon fixation by photosynthesis.

This basic ecological testing should give an indication of the type of ecological damage that may occur. These tests should cover, as far as possible, primary ecological functions and widely differing species. Further testing may be desirable for several reasons, for example:

a.  high potential for bioaccumulation
b.  low degradability
c.  production, use and disposal patterns indicating a potential ecohazard
d.  high or highly specific (eco)toxicity.

Decisions on any further testing should be taken in the context of all evidence available, including that from adjacent fields of research as defined in the relevant chapters of the OECD Guidelines. An indication of possible "triggers" for further testing is shown in Figure 4.

The OECD-scheme of testing is still under development, especially with regard to the later phases of testing and to non-aquatic tests. The scheme given above must therefore only be seen as a possible way of using these guidelines and has no official status. Nevertheless we may accept the OECD-scheme as a well-considered effort to harmonize testing in a number of countries, and to create some order in a confusing situation.

In the context of the subject of the present publication, some obvious shortcomings of the scheme resulting from its limited objectives (viz. to provide a system of tests for the prediction of possible harm from new and existing chemicals) should be mentioned. The tests are intended for pure or well defined chemicals. Mixtures of chemicals occuring in polluted situations, require adaptation of the specified methods. The philosophy outlined above, however, remains valid. Furthermore, the scheme was not devised to cope with the diversity of situations in actual cases of pollution where, for example, synergism with pre-existing pollutants may occur and wehere physico-chemical conditions may differ greatly from laboratory conditions. Moreover, the interaction with co-existing populations of organisms is difficult to handle. Last not least it is apparent that only few organisms lend themselves easiliy to use in testing. A validation of the predictions of this scheme of prospective testing of chemicals as to its per-

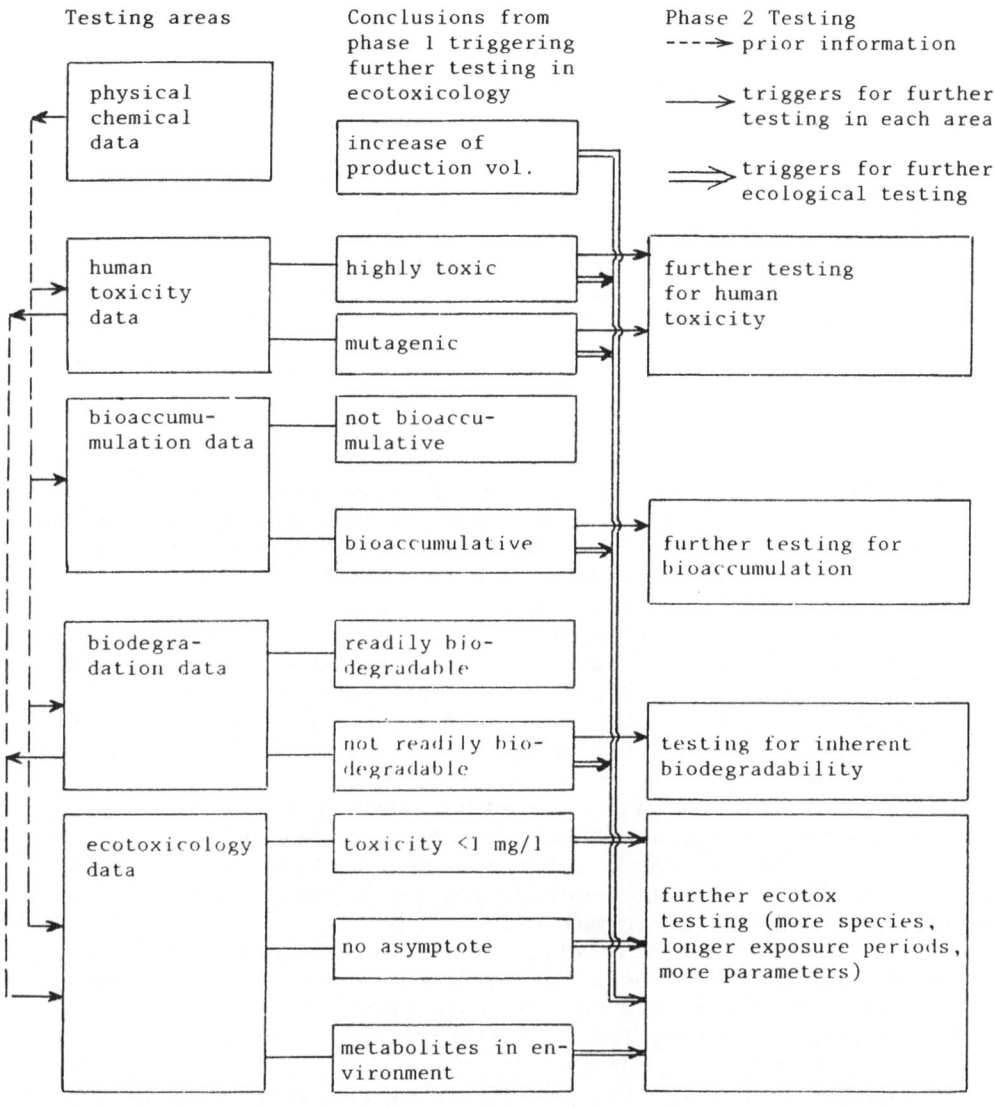

From Anonymous 1979

Figure 4: The relations between the triggers from various fields
of the assessment of potential environmental effects
leading to further ecotoxicolgical testing.

formance in actual situations therefore is very much needed. The analysis of ecoaccidents offers such an opportunity, provided that the inference drawn from these accidents are such that they can be used for comparative purposes. A few considerations on such comparisons is given in the following chapter.

THE PROSPECTIVE TESTING OF CHEMICALS AND ECOACCIDENTS

In an ecoaccident we are dealing with a situation which fulfills the requirement of the first of the Kochian principles, outlined in the introduction, viz. the presence of a suspected harmful agent in such an amount and distribution that it might explain some untoward phenomena observed at the place of the accident. However, one must be careful in attributing cause-effect relationships in such a situation, when only a correlation is apparent. Bowman (1978) has given a good example of the necessity of additional research to find the real cause-effect relation; In this case a small oil spill was associated with the presence on the coast of many dead oil-contaminated barnacles. It could, however, be shown that only part of the mortality of the barnacles was directly due to smothering with oil from the oil spill. The greater part of the mortality was due to "natural causes" also observed in adjacent non-oiled areas. Moreover it was clear that the direct effect of oil should have been distinguished from the action of the oil-disperant used for cleaning the site. Full advantage of the experience with ecoaccidents can therefore, only be gained if research is not restricted to observations on the site of the accident itself. These observations should be complemented with a number of tests as outlined in the previous chapters. In making a choice of appropriate tests, the diagram in the introduction must be considered: we are dealing with the dose of a certain chemical coming in contact with a certain receptor (biotic system) in which certain effects are then made manifest. This whole chain of events occurs under certain environmental conditions. The following requirements for a valid assessment of the case may therefor be formulated.
1.  To deal with the scientific analysis of the accident (as distinct from the technical handling of the cleaning operation) expertise should be available from:
    - analytical chemistry for definition of the agent and its dose
    - taxonomy for definition of the affected organisms
    - ecology for definition of the effects on the site, and for description of the environment.
2.  A sample of the agent (presumably a chemical, or mixtures of chemicals) should be collected. It should be analyzed and

defined as to possible contituent active agents according to the second of the (adapted) Kochian principles referred to in the introduction. The sample should be kept for future reference.

3. The dose and the distribution of the agent at the site of the accident should be estimated according to the first of the Kochian principles.

4. The affected organisms at the site of the accident should be sampled and be preserved for further reference. Molluses are appropriate accumulators of lipophilic compounds, whereas Seaweeds and the like may be used for the same purpose with hydrophilic compounds.

5. An ecological survey of possible effects at the site of the accident should be made in comparison with a blank situation. The blank situation may be available in an adjacent non-affected area or from historical records. This requirement is imperative for compliance with the first Kochian principle.

6. The observed effects should be simulated with the isolated agent or agents under controlled conditions. Ideally this should be carried out with the organisms sampled on the spot, but compromises including the use of similar organisms must be expected. This requirement is essential for complying with the third Kochian principle.

7. For a complete ecological survey it is desirable, if possible, to observe the restoration of the damaged ecosystem, if any.

8. If the validation of prospective tests, is also desired, a full array of tests such as provided in the OECD-Guidelines should be carried out.

Last not least, one may wonder whether the procedure outlined would contribute to dealing technically with the consequences of ecoaccidents. The answer must be in the affirmative since effective control of untoward effects observed requires a thorough knowledge of the causative agent. Mixtures of chemicals, or simultaneous presence of active agents (e.g. oil spills treated with emulsifiers) are often involved; the identification of the causative agent in these cases is of paramount importance.

SUMMARY

1. This paper gives an outline of a scheme of prospective tests for the prediction of environmental effects of chemicals.

2. It is pointed out that application of the Kochian principles, adapted to the requirements of ecotoxicology, is imperative for the retrospective analysis of field experiments, including ecoaccidents.

3. A brief description of the principle of the scheme of tests laid down in the "OECD-Guidelines" is given.
4. Recommendations for the application of such a prospective test scheme in actual cases of ecoaccidents are also given.


ACKNOWLEDGMENT

The philosophy of testing as presented briefly in this paper was first formulated by Hueck & Hueck-van der Plas (1976) and further developed in Hueck (1978) and Hueck-van der Plas & Hueck (1979). These publications formed the basis for discussion in the "OECD Expert Group on Ecotoxicology" in which further contributions were made by a number of experts as recorded in Anonymous (OECD), 1981.
Quotations from these publications were used in the present paper. They are a personal choice of the author and do not represent any official opinion.


LITERATURE

Anonymous (1972). "Convention for the Prevention of Marine Pollution by Dumping from ships and Aircraft", Oslo 15 February 1972. Ref. Traktatenblad 6 June 1972. Government Printing Office The Hague, The Netherlands.

Anonymous (1976). "Proposal for a Council Directive the Sixth modeferation of the Council Directive of dangerous substances". Official Journal of the European Communities No C 260 (4), page 270 - 285 (5 November 1976).

Anonymous (1979). "Report on the assessment of potential environmental effects of chemicals; the effect on organisms other than man and on ecosystems". 3 Volumes. A report to the Organisation for Economic Cooperation and Development (OECD). Environmental Committee - Chemicals Group - Chemicals Testing Programme. Paris, France. Prepared by "Division of Technology for Society TNO Department of Biology" & "Study and Information Centre TNO for Environmental Research Delft, The Netherlands" on behalf of the "OECD, Chemicals Testing Programme Ecotoxicology Group".

Anonymous (1980). "Degradability, Ecotoxicity and Bio-Accumulation". The determination of the possible effects of chemicals and wastes on the aquatic environment. Government Publishing Office, The Hague, The Netherlands.

Anonymous (1981). "OECD Guidelines for Testing of Chemicals". Organisation for Economic Cooperation and Development. 2, Rue Andre Pascal, 75775 Paris CEDEX 16, France.

Bowman, R.S. (1978). "Dounray Oil Spill: Major Implications of a Minor Incident". Marine Pollution Bulletin 9, 269-273.

Hueck, H.J. (1978). "Principles of Testing for Potential Biotic Environmental Effects of Chemicals". Berichte Umwelt Bundes Amt (Erich Schmidt Verlag, Berlin) 10, 23-37.

Hueck, H.J. & H. Hueck-van der Plas (1976). "A system of tests for the assessment of potential effects of chemicals and quality data in the environment." in R. Amavis & J. Smeets (eds.). Principles and methods for determining ecological criteria on hydrobiocenoses, p. 468-528. Published for the Commission of the European Communities by Pergamon Press, Oxford U.K.

Hueck-van der Plas, E.H. & H.J. Hueck (1979) "The prospective assessment of Environmental Effects of Chemicals". Drug Design 8, 311-354.

Koch, R. (1882). "Über die Milzbrandimpfung" Verlag Theodor Fischer, Kassel und Berlin.

Lindeman, R.L. (1942). "The trophic-dynamic aspect of ecology" Ecology 23 399-418.

Warren, C.E. (1971). "Biology and Water Pollution Control". W.B. Saunders Comp. Philadelphia U.S.A.

FUTURE NEEDS

John Cairns, Jr.

University Center for Environmental Studies and
Department of Biology
Virginia Polytechnic Institute and State University
Blacksburg, Virginia   24061 USA

Although a certain amount of predictive modeling can be carried
out in experimental systems, particularly multispecies toxicity
tests, microcosms, mesocosms, and field enclosures, case histories
of ecoaccidents are the best source of information for modeling.
More creative use must be made of these episodic events, particularly
with regard to mitigation of the short-term deleterious effects
and rehabilitation of the system once the perturbation has been
adequately diminished or has disappeared.  Although ecoaccidents
are unfortunate, using them to carry out studies that will improve
management and mitigation practices for future accidents is essential.
For example, several clean-up techniques could be used simultaneously
in ecologically comparable areas for some spills, and the immediate
and long-term effects of each practice could be better documented.
Similarly, recolonization of perturbed areas following various
clean-up techniques could be studied, and the effect of different
practices would then be quantitatively documented.  If we are to
optimize the information gathered, a greater flexibility must exist
between regulatory agencies, the industry or other organization
responsible for the ecoaccident, and the general public.  The amount
of information available upon which to base management practices
following ecoaccidents is not adequate for a "cookbook" approach
for most accidents.  In cases where adequate knowledge is available,
ecoaccidents can be treated in a fairly routine manner.  Where it
is not, the follow-up actions should optimize both the gathering
of additional information by using the ecoaccident as an opportunity
for research and for validation of predictive models already
developed on other sites.  In order to make effective use of
ecoaccidents, some organization must accept the responsibility
of identifying research and informational needs, determining how

151

these might best be acquired and the circumstances that will give the most useful information, and developing a plan that can be immediately implemented when the opportunity arises. The period following an ecoaccident is neither the ideal time for lengthy negotiations between regulatory agencies and industries nor a suitable time to develop a research plan de novo. Established research plans can be modified to fit a particular site and set of circumstances when the emergency occurs. The basic plan and its scientific credibility should be thought out well in advance of the ecoaccident.

An international clearinghouse for such information is most desirable. Such organizations as the Union Internationale des Sciences Biologiques or the Man and the Biosphere Program might well serve this purpose. A small number of countries might also set up ecoaccident teams under the auspices of such groups as the Scientific Branch of the North Atlantic Treaty Organization. These would visit the site of an ecoaccident either in the territory of a participating country or when caused by an accident of a ship, airplane, etc. belonging to one of the NATO countries.

Also needed is an array of ecological preserves designed to furnish colonizing species for damaged areas. These ecological preserves might be used in a number of ways: (1) as a source of "pedigreed" species for restocking damaged ecosystems, (2) for multispecies, micro- and mesocosm studies, and (3) as baseline ecosystems or reference ecosystems against which rehabilitation of damaged ecosystems can be compared. The concept of biological reserves is not new since UNESCO has approximately 30 such sites (Franklin, 1977). Project AQUA of the International Biological Program established a large number of reserves throughout the world (Luther and Rzoska, 1971). Within these baseline ecosystems, studies should include: (1) understanding the dynamics of ecosystems, (2) developing mathematical models to facilitate understanding of the interactions of components within each system, and (3) using field studies to determine the degree of information redundancy in data on both the structure and the function of these systems. The latter would make sampling following an ecoaccident more cost effective and more informative.

Because of the paucity of individuals trained in ecotoxicology, implementation of many of these recommendations, particularly those calling for site-specific research following an ecoaccident, will be exceedingly difficult. Classical ecologists do not know enough toxicology to cope effectively with an ecoaccident. Furthermore, they may often not be accustomed to working under the emergency procedures required in such circumstances, particularly when a number of other disciplines is involved. To remedy this situation, more funds should be made available to support the education and training of students in ecotoxicology at all levels.

The two areas that can be addressed most quickly, most effectively, and with the most immediate benefits are: (1) steps to reduce the number of ecoaccidents due to the various factors listed in Baker's chapter (this volume) and elsewhere in this book, and (2) more effective exchange of information following an ecoaccident coupled with the better coordination of the various organizations responsible for coping with it. These are not scientific matters, but they will enormously enhance the effectiveness of the scientific component of the ecoaccident problem solving group. Additionally, there will be fewer damaged ecosystems.

LITERATURE CITED

Franklin, J. J., 1977, The biosphere reserve program in the United States, Science, 195:262.
Luther, H., and Rzoska, J., 1971, "International Biology Program Handbook #21," Blackwell Scientific, Oxford.

CONTRIBUTORS

Baker, J., Dr.
Field Studies Council
Preston Montford
Shrewsbury, SY4 1HW
United Kingdom

Bro-Rasmussen, F., Dr.
Laboratory of Environmental
    Science and Ecology
Technical University of Denmark
Building 224
DK 2800 Lyngby, Denmark

Calamari, D., Dr.
Institute of Entomology
Faculty of Agriculture
University of Milan
Via Celoria, 2 20133
Milano, Italy

Cattabeni, F., Dr.
Institute of Pharmacology
    and Pharmacognosy
University of Milan
Via A. Del Sarto 21
20129 Milano, Italy

Hueck, H. J., Dr.
Division of Technology for Society
TNO
P.O. Box 217
2600 AE Delft
The Netherlands

Osborn, D., Dr.
NERC, Institute of Terrestrial
    Ecology
Monks Wood Experimental Station
Huntingdon, CAMBS PE17 2LS
United Kingdom

Peter, H. W., Dr.                    Umweltbundesamt
                                     Bismarckplatz 1
                                     D-1000 Berlin 33
                                     Federal Republic of Germany

Tiravanti, G., Dr.                   Water Research Institute (IRSA)
Passino, R.                          National Research Council (CNR)
                                     Rome, Italy

Warnøe, K., Ms.                      Danish National Agency of
                                        Environmental Protection
                                     Strandgode
                                     DK-1401 Copenhagen, Denmark

PARTICIPANTS IN THE FIRST PREPARATORY MEETING ON <u>ECOACCIDENTS</u>
31 March - 1 April 1980


Prof. Dr. D. J. Kuenen
Chairman

Emeritus Professor
Leiden University
Blauwe Vogelweg 2a
2333 VK Leiden
The Netherlands

Ms. Drs. E.H. Hueck-van der Plas
Secretary

Study and Information Centre TNO
    on Environmental Research
Professor Schoemakerstraat 97
P. O. Box 186, 2600 AD Delft
The Netherlands

Ms. Dr. J. Baker

Head, Oil Pollution Research Unit
Orielton Field Centre
Pembroke, Pembrokeshire
United Kingdom

Dr. J. A. B. Barstad

Head, Department of Environmental
    Toxicology
National Institute of Public
    Health
Geitsnyrsveien 75, Postuttak
Oslo 1, Norway

Dr. J. Burton

Institute of Environmental
    Studies
University of Toronto
Ontario, Canada

Dr. D. Calamari

Istituto di Ricercha sulle Acqua
Reparto Sperimentale
    Idrobiologica Applicata
Via Occiate, Brugherio
Milano, Italia

Dr. R. Harris

National Aeronautics and
    Space Administration
Langley Research Centre
Hampton, Virginia 23665 USA

Dr. H. J. Hueck

Head, Department of Biology
Division of Technology for
    Society TNO
Schoemakerstraat 97
P. O. Box 217
2600 AE Delft
The Netherlands

Dr. Ir. N. van Lookeren Campagne

Shell Nederland BV
Hofplein, Rotterdam
The Netherlands

PARTICIPANTS IN THE SECOND PREPARATORY MEETING ON <u>ECOACCIDENTS</u>
14-15 October 1982


Prof. Dr. D. J. Kuenen
Chairman

Council for Environmental
    and Nature Research
Huis te Lande laan 492
2283 VJ Rijswijk (ZH)
The Netherlands

Ms. Drs. E.H. Hueck-van der Plas
Secretary

Study and Information Centre TNO
    on Environmental Research
Professor Schoemakerstraat 97
P. O. 186, 2600 AD Delft
The Netherlands

Dr. A. L. Buikema, Jr.

University Center for
    Environmental Studies
Virginia Polytechnic Institute
    and State University
Blacksburg, Virginia 24061 USA

Dr. R. Cabridenc

Institut National de Recherche
    Chemique Appliquee (IRCHA)
Centre de Recherche
Boite Postale No. 1
91710 Vert le Petit, France

Prof. Dr. J. Cairns, Jr.

University Center for
    Environmental Studies
Virginia Polytecnic Institute
    and State University
Blacksburg, Virginia 24061 USA

Prof. Dr. D. Calamari

Istituto di Ricercha sulle Acqua
Consiglio Nazionale delle
    Richerche
Via Occiate, 20047 Brugherio
Milano, Italia

Dr. Brian Dicks

Field Studies Council
Orielton Field Centre
Pembroke DYFED SA 71 5EZ
Great Britain

Dr. J. van Eyk

Koninklijke Shell Laboratorium
Badhuisweg 3,
1031 CM Amsterdam
The Netherlands

Dr. H. J. Hueck

Senior Scientist
Division of Technology for
    Society TNO
P. O. Box 217, 2600 AE Delft
The Netherlands

Ir. J. L. Nooyen

Koninklijke Shell Laboratorium
Badhuisweg 3,
1031 CM Amsterdam
The Netherlands

Dr. H. W. Peter

Umweltbundesambt
Bismarckplatz 1,
1000 Berlin 33
Budesrepublik Deutschland

INDEX

Tetramethyl lead (continued)
  concentration in sea water,
      36, 37
  toxicity of, 27
Toxicity tests, 81-88, 120-130,
      135-140, 142, 143, 145,
      151
Toxic Substances Control Act,
      98, 117
Trichloroethylene
  bacteria indication test, 125
  biodegradation of, 122, 123,
      127
  carcinogenicity of, 123, 124
  characteristics of, 119
  ecotoxicological studies of,
      118, 121-122, 123, 125,
      129
  long-term effects, 128
  mutagenicity of, 123, 128
  physico-chemical properties of,
      120, 125
  teratological study of, 123
  toxicological studies of,
      120-121, 122, 127-129

Wading birds (*see also* Dunlin)
  lead concentration in, 71
  mortality of, 59, 73, 74
  toxic chemical concentration
      in, 71
Wildfowl
  lead concentration in, 59, 60